Linearization and Efficiency Enhancement Techniques for Silicon Power Amplifiers

Linearization and Efficiency Enhancement Techniques for Silicon Power Amplifiers

From RF to mmW

Eric Kerhervé
University of Bordeaux, IMS Laboratory, CNRS UMR 5218,
Bordeaux INP France

Didier Belot
STMicroelectronics, Rue Jean Monnet, Crolles;
CEA-LETI, Rue de Martyrs, Grenoble

AMSTERDAM • BOSTON • HEIDELBERG • LONDON
NEW YORK • OXFORD • PARIS • SAN DIEGO
SAN FRANCISCO • SINGAPORE • SYDNEY • TOKYO

Academic Press is an imprint of Elsevier

Academic Press is an imprint of Elsevier
125, London Wall, EC2Y 5AS
525 B Street, Suite 1800, San Diego, CA 92101-4495, USA
225 Wyman Street, Waltham, MA 02451, USA
The Boulevard, Langford Lane, Kidlington, Oxford OX5 1GB, UK

Notices
Knowledge and best practice in this field are constantly changing. As new research and
experience broaden our understanding, changes in research methods or professional practices,
may become necessary.

Practitioners and researchers must always rely on their own experience and knowledge in
evaluating and using any information or methods described herein. In using such information
or methods they should be mindful of their own safety and the safety of others, including
parties for whom they have a professional responsibility.

To the fullest extent of the law, neither the Publisher nor the authors, contributors, or editors,
assume any liability for any injury and/or damage to persons or property as a matter of
products liability, negligence or otherwise, or from any use or operation of any methods,
products, instructions, or ideas contained in the material herein.

ISBN: 978-0-12-418678-1

Library of Congress Cataloging-in-Publication Data
A catalog record for this book is available from the Library of Congress

British Library Cataloguing-in-Publication Data
A catalogue record for this book is available from the British Library

For Information on all Academic Press publications
visit our website at http://store.elsevier.com/

This book has been manufactured using Print On Demand technology.

Working together
to grow libraries in
developing countries

www.elsevier.com • www.bookaid.org

CONTENTS

LIST OF CONTRIBUTORS

Mohammadhassan Akbarpour

Intelligent RF Radio Laboratory (iRadio Lab), Department of Electrical and Computer Engineering, University of Calgary, Calgary, AB, Canada T2N 1N4

Alexis Aulery

University of Bordeaux, IMS Laboratory, CNRS UMR 5218, Bordeaux INP France

Didier Belot

STMicroelectronics, Rue Jean Monnet, Crolles; CEA-LETI, Rue de Martyrs, Grenoble

Dominique Dallet

University of Bordeaux, IMS Laboratory, CNRS UMR 5218, Bordeaux INP France

Kaushik Dasgupta

California Institute of Technology (Caltech), Pasadena, CA

Nicolas Delaunay

University of Bordeaux, IMS Laboratory, CNRS UMR 5218, Bordeaux INP France; STMicroelectronics, Rue Jean Monnet, Crolles

Nathalie Deltimple

University of Bordeaux, IMS Laboratory, CNRS UMR 5218, Bordeaux INP France

Fadhel M. Ghannouchi

Intelligent RF Radio Laboratory (iRadio Lab), Department of Electrical and Computer Engineering, University of Calgary, Calgary, AB, Canada T2N 1N4

Ali Hajimiri

California Institute of Technology (Caltech), Pasadena, CA

Mohamed Helaoui

Intelligent RF Radio Laboratory (iRadio Lab), Department of Electrical and Computer Engineering, University of Calgary, Calgary, AB, Canada T2N 1N4

Eric Kerhervé

University of Bordeaux, IMS Laboratory, CNRS UMR 5218, Bordeaux INP France

Bertrand Le Gal

University of Bordeaux, IMS Laboratory, CNRS UMR 5218, Bordeaux INP France

Baudouin Martineau

STMicroelectronics, Crolles, France; Univ. Grenoble Alpes, F-38000 Grenoble, France; CEA, LETI, MINATEC Campus, F-38054 Grenoble, France

Earl McCune

RF Communications Consulting, Santa Clara, CA

Ullrich R. Pfeiffer

High-Frequency and Communication Technology, University of Wuppertal, Wuppertal, Germany

Chiheb Rebai

GRES'COM Lab, SUP'COM, University of Carthage, Tunisia

Neelanjan Sarmah

High-Frequency and Communication Technology, University of Wuppertal, Wuppertal, Germany

CHAPTER *1*

Holistic Approaches for Power Generation, Linearization, and Radiation in CMOS

Ali Hajimiri and Kaushik Dasgupta
California Institute of Technology (Caltech), Pasadena, CA

The field of wireless communications has experienced an exponential growth over the past four decades, going from the realm of science fiction to becoming so ubiquitous and natural that the younger generations have a difficult time imagining a world without it. This has been made possible through many breakthroughs in our understanding of the nature of wireless communications and, more importantly, numerous innovative enabling technologies that have made personal wireless communication an everyday reality.

Creating the fascinating edifice that is the connected world of ubiquitous access to information has been made possible through effective complexity management performed through a process of divide and conquer. This is possible through a systematic process of specialization and subspecialization in electrical engineering and its associated fields that have been part of the historic trends in this area, creating levels of abstractions within which a small group of people can design for a given set of specifications. This is what has enabled this complexity management. These levels of abstraction have played a key role in our ability to deploy such complex systems such as our wireless communication systems. A typical example of the hierarchy of these levels of abstractions is shown in Figure 1.1.

In radiofrequency integrated circuit (RFIC) design, the same trends have been present since the inception of RFIC and monolithic microwave integrated circuits (MMICs). The system architectures are defined and created at a separate time and by a different group of engineers than those who define the chip architectures and those who

Linearization and Efficiency Enhancement Techniques for Silicon Power Amplifiers.
DOI: http://dx.doi.org/10.1016/B978-0-12-418678-1.00001-5

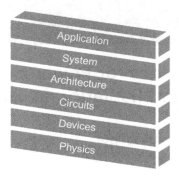

Figure 1.1 Levels of abstraction in a system.

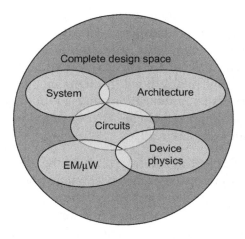

Figure 1.2 The design space and its subspaces, limiting the range of design possibilities considered.

design the transistor levels circuit blocks. The electromagnetic design (e.g., antenna and off-chip passive components) has also been generally performed by yet another group of designers. This partitioning of tasks has enabled speedy and tractable design of such systems; however, it has also created impediments in the way of innovation by limiting the number of possibilities that can be considered in the design of the entire system by constraining the design space to a limited subset of all possibilities (Figure 1.2).

These somewhat arbitrary levels of abstractions have resulted in some of the classical approaches to RFIC design, with the individual circuit elements considered as lumped components and the effects of nonidealities and the couplings modeled using additional parasitic

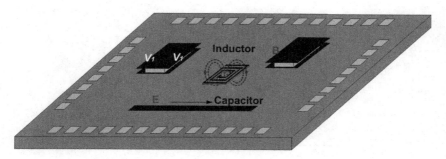

Figure 1.3 A classical RFIC integrated circuit.

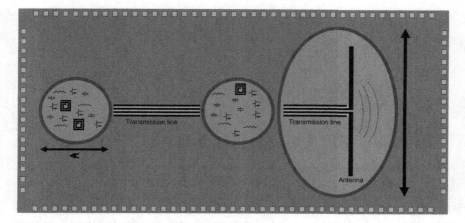

Figure 1.4 A standard MMIC operating at high frequencies.

components. This allows for continuous application of long-known circuit simulation techniques essentially based on nodal analysis (Figure 1.3). Even in the realm of MMIC, the approach has been only partially extended by analyzing the nonlumped blocks such as transmission lines and on-chip antennas as separate units and representing them as scatter parameter boxes (Figure 1.4).

However, the ever-increasing complexity of wireless systems and the associated integrated circuits combined with the constantly growing standards and frequency bands of operations have created strong interconnections among these levels of abstractions. This is a challenge if examined from the point of view of classical levels of abstractions. It manifests itself as a cohort of problems such as cross talk, electromagnetic coupling, impedance mismatches, parasitic elements, and others. This situation has been exacerbated by the scaling

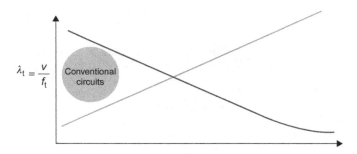

Figure 1.5 The conceptual plot of the cutoff wavelengths and chip dimensions versus time.

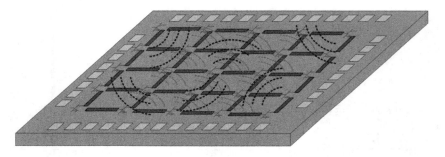

Figure 1.6 Highly parallel, strongly coupled holistic integrated EM structures.

of the transistors; the cutoff wavelength (the wavelength associated with the cutoff frequency of the transistors) is constantly shrinking and the dimensions of the integrated circuits have been growing, as conceptually shown in Figure 1.5. This has resulted in a crossing of the two curves (which happens approximately at the 250-nm CMOS nodes), leaving us in a nonlumped regime.

However, if one is willing to look at the problem in a broader context and not be confined by the classical levels of abstraction, then this can present a tremendous opportunity to overcome some of the classical problems in RFIC design. This can be accomplished by using a holistic approach to co-design of various parts of the system [1,2]. The holistic approach relies on a more close interaction of the electromagnetics and the transistors. It generally achieves this through a large number of transistors and small passive structures with strong electromagnetic coupling operating in concert (Figure 1.6).

These structures can be used to enable simultaneous and active control of the electric and magnetic field profiles (E and H), enabling a much richer combination of field profiles and a much broader set of functions. This is an example of holistic integration of architecture, circuits, electromagnetics, and devices.

Although promising, such holistic approaches require the designers to obtain a broader set of skills ranging from a more thorough understanding of the system level matters to deeper device and physics-oriented aspects of the system. However, once the walls between classical levels of abstraction are removed, it becomes possible to design circuits that can outperform the classical solutions. More sophisticated approaches to simulation of these systems, which necessitate rethinking of the whole design flow, are also necessary.

A holistic approach inevitably leads to several underlying trends in the design methodology. Some of those trends are:

1. *Parallelism*: One of the most significant features of today's silicon integrated circuits is the practically unlimited number of transistors they offer. As integrated circuits find their way into every conceivable part of wireless systems, it becomes essential to take full advantage of the large number of transistors. This unlimited number of transistors, however, comes at the cost of more limited power-handling capability of individual ones. This makes electromagnetically parallel structures much more compelling and a necessity in these approaches. This mindset can be loosely stated as "the army of mice versus the giant elephant."

2. *Concurrency*: While related to parallelism at some level, concurrency in this context refers to various signal paths that process potentially different pieces of information in parallel.

3. *Highly reconfigurable, modifiable, and healable*: To be able to address various applications and take full advantage of the co-integration at various levels of abstraction, highly reconfigurable and dynamically modifiable systems are essential. In the presence of a large number of elements involved, this reconfigurability must be enhanced and automated, giving rise to so-called self-healing systems. Such self-healing systems are highly conducive to holistic approaches.

We discuss these themes through a few examples in the remainder of this chapter.

1.1 SELF-HEALING INTEGRATED CIRCUITS

The past few decades have witnessed aggressive CMOS scaling, primarily driven by the demand for cost-, area-, and power-efficient processors for desktops as well as mobile chipsets. The number of transistors per chip has increased from a few thousand in the 1980s to a few billion in the latest desktop processors. More recently, pure dimensional scaling has become less effective and other performance enhancement techniques like strained Si [3,4], tri-gate [5] FinFETs, and others have been widely adopted by the semiconductor industry. While bringing significant improvements in power consumption, area, and overall performance, such scaling also comes at the cost of increased variations—both between chips and on the same chip. We can broadly categorize these variations into static variations and dynamic variations. Static variations can mostly be attributed to the fabrication process itself and is dominated by two major effects—random dopant fluctuations (RDFs) and line edge roughness (LER).

As shown in Figure 1.7A, the average number of dopant atoms in the channel region has scaled down dramatically, from a few hundred in 130-nm CMOS to less than one hundred in modern CMOS transistors. Any variation in both the number and the actual placement of these dopants (RDF) thereby leads to significant changes in the transistor performance. LER is caused by imperfections during the lithographic process itself, which directly affects device overlap capacitances as well as other critical device performance parameters. Figure 1.7B shows how the threshold voltage variation scales with process nodes and the contributions of RDF and LER.

In addition to these static variations, changes in the operating conditions of a CMOS IC can contribute to performance degradation over time. These dynamic variations typically include thermal changes, transistor aging, and (in the specific case of PAs) antenna impedance mismatch caused by changes in the near field of the antenna. Another issue typically plaguing RF designers is the lack of reliable high-frequency models for these deeply scaled CMOS devices. This is due to the fact that the main driving force behind CMOS scaling has been the digital processor market, and thus most commercial foundry models are optimized for digital design and are often unpredictable at higher frequencies.

There are two different approaches to tackle these variations. The first approach is to design systems that are inherently more robust by

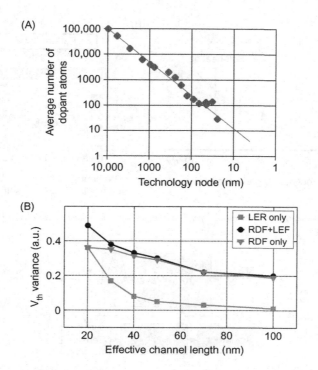

Figure 1.7 (A) Average number of dopant atoms scaling with technology [6] *and (B) various components of threshold voltage variation.*

utilizing design techniques and circuit topologies less sensitive to process and mismatch variation. This approach is feasible at lower frequencies for which methods like supply voltage optimization, optimum device sizing, and others have been widely adopted. However, at mm-wave and RF frequencies, their application has been limited primarily due to the negative impact these techniques generally have on the overall circuit performance. A much more scalable and systematic approach is self-healing and draws its inspiration from nature's own way of countering adverse environmental effects. In a self-healing system, the CMOS IC dynamically senses the performance degradation and actuates itself back to its optimum performance by adjusting various knobs in the circuit. The goal of self-healing is thus two-fold. First, it seeks to recover the performance advantages of CMOS scaling by making the entire design less sensitive to modeling inaccuracies. Second, it improves the yield of such systems, thereby reducing overall cost as shown in Figure 1.8.

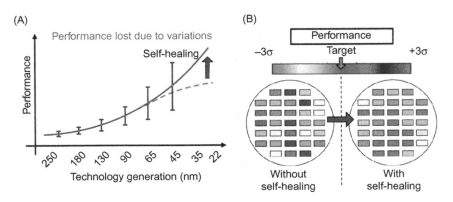

Figure 1.8 Self-healing: (A) improving performance as well as reducing variations and (B) yield improvement [7].

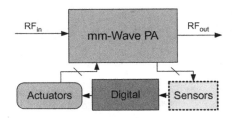

Figure 1.9 Block diagram of a generic self-healing mm-wave PA.

This technique takes advantage of the vast digital processing power commonly available in modern CMOS processes through an on-chip feedback loop involving sensors, actuators, and a digital optimization component (Figure 1.9).

Depending on the actual design, a variety of design constraints are applicable to the building blocks of a self-healing system. First, the sensors need to be robust enough for the same variations they are trying to sense. Because these sensor circuits do not necessarily need to be optimized for performance, this robustness can simply be achieved by using variation-tolerant topologies/methodologies. In most self-healing systems, the optimization/algorithm component is implemented in the digital domain and, as such, these analog sensors need to interface with the digital core using on-chip analog-to-digital converters (ADCs). Regarding the actuators, care must be taken such that they cover a wide enough actuation space to allow the system to heal against a wide variety of sources of variation, some of which may not have been anticipated during the design phases.

1.1.1 Self-Healing mm-Wave Power Amplifier

The advantages of self-healing are discussed in the context of a fully integrated self-healing PA at 28 GHz implemented in a 45-nm SOI CMOS process [8]. The core architecture of the PA is based on a two-stage two-to-one power combining class AB PA where the input and output are matched to 50 Ω. The PA is designed for maximum saturated output power where the driver stages are sized to be able to drive the output stages fully into saturation. Figure 1.10 shows the architecture of the PA along with the associated blocks enabling self-healing.

Each power stage is designed as a cascode amplifier for higher gain. The cascode transistors are thick-oxide transistors able to handle the large voltage swings at the output. For a PA, the typical performance metrics of interest are output power, gain, and efficiency, meaning the sensors need to be able to estimate input and output power as well as DC current drawn by the PA. Coupled line-based RF sensors are placed at the input and output ports of the PA, and voltage regulator-based DC sensors sense both the PA and driver currents. Regarding actuators, gate bias actuators control quiescent operating points of all transistors and tunable transmission line actuators provide an added degree of freedom with respect to reconfigurability of the output matching network.

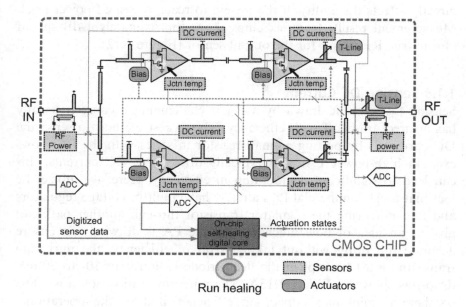

Figure 1.10 Architecture of the example integrated self-healing PA [8].

1.1.2 Sensing RF Power

RF power can be sensed by simply using a voltage sensor at the output as long as the load remains constant (typically 50 Ω). However, in the presence of load variations due to Voltage Standing Wave Ratio (VSWR) events, this estimate may be completely off. In this example PA, coupled transmission lines are used to sense input and output power. The coupling ratio is kept small enough to avoid significant performance degradation of the amplifier. Because of this small coupling ratio, the coupled line segments can be much shorter than a quarter wave; in this example, 220-μm couplers with a coupling ration of ∼ 20 dB have been used. The couplers are designed for a characteristic impedance of 50 Ω and are terminated by the same characteristic impedance at the coupled and isolated ports. Figure 1.11 shows the structures of the input and output couplers along with the measured S-parameters [9].

The output voltage across the 50-Ω resistors are then rectified using MOS-based transistors biased at cutoff, filtered, amplified, and finally converted to an analog DC voltage. In addition, by sensing both through and isolated powers across the coupled line, it is possible to estimate the real power delivered to a load in the presence of load mismatch. We return to this concept in the measurement section. Variations in power sensing is also of prime importance because it directly affects the ability of the system to reach a desired power level. Measurement results from six chips show approximately 1-dB spread (3σ) in true RF power for the output sensor (Figure 1.12).

1.1.3 Sensing DC Current

Sensing DC current drawn by a PA is challenging because it usually has a direct trade-off with efficiency. The easiest method to sense the DC current is by putting a small resistor in series with the PA; however, in high-power PAs that typically draw large DC currents, this can lead to significant degradation in efficiency. Here, we utilize the fact that most commercial PAs already have built-in voltage regulators and that mirroring the regulator transistor through another path can give us an idea of how much current the PA is drawing. To ensure accurate mirroring without taking a hit in efficiency, the mirroring transistors must be kept in the deep triode region with 10- to 30-mV drops, as shown in Figure 1.13A. The mirroring ratio chosen is 100:1 so there is minimum impact on efficiency and so the operational

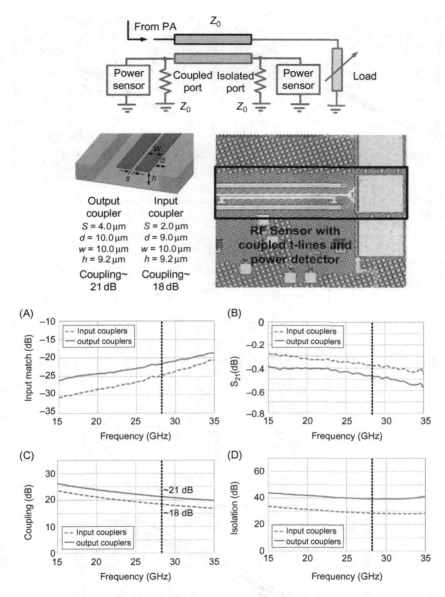

Figure 1.11 Dimensions of the input and output couplers as well as S-parameter measurements [9].

amplifier A_1 ensures accurate mirroring (Figure 1.13A). Measurement results in Figure 1.13B show variation in the DC sensor measured over five chips. The 3σ spread of DC current based on a particular sensor reading is approximately 14 mA.

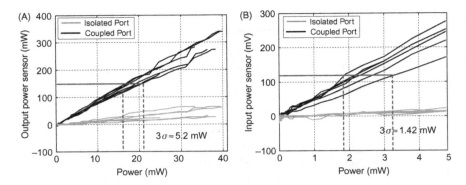

Figure 1.12 Measured RF power sensor response for (A) output and (B) input sensors at 28 GHz [8].

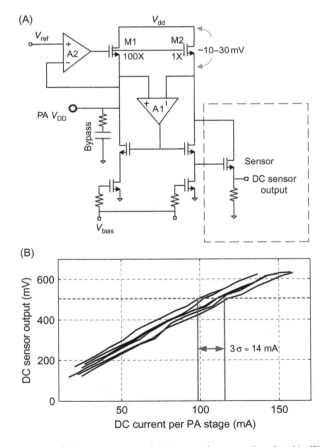

Figure 1.13 (A) Schematic of DC current sensor and (B) measured responses from five chips [8].

1.1.4 Actuating Quiescent Operating Point

In general, the PA must be biased at a current that maximizes f_{max} of the transistor. However, as seen in the previous sections, the threshold voltage variation can easily cause nonoptimal operation of the amplifier, leading to reduced output power and efficiency. The simplest way to actuate the operating point of the transistor is to adjust gate voltages of the transistors in both the driver and output stages. Such actuators can be very low overhead because of the high DC gate resistance of CMOS transistors. These actuators can also be used to control the gate bias in back-off power levels, thereby improving drain efficiency. In the example PA, the gate bias actuation is implemented through low-power digital-to-analog converters (DACs) at the gates of the common source and the cascode transistors of both the driver as well as the output stage. These DACs are implemented as binary-weighted current mode DACs drawing less than $500\,\mu W$ per DAC under nominal conditions. Figure 1.14 shows measurement results for the implemented DACs.

A second type of actuator implemented in this example PA is passive matching network actuators. These actuators allow us to tune the output matching network of the PA after fabrication to counter variations in parasitic capacitances and to ensure optimal match under process variations. In addition, in the presence of antenna VSWR events induced by changes in the operating environment of the PA, these actuators can dynamically adjust the match, ensuring optimum operation under a wide variety of load mismatch conditions. In the present design, tunable transmission line stubs have been utilized that allow us

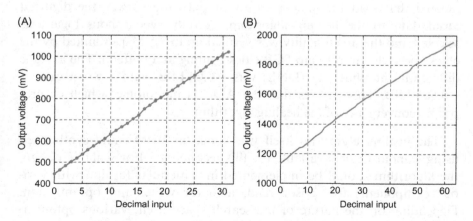

Figure 1.14 Measurement results from DACs controlling (A) common source and (B) cascode transistor [8].

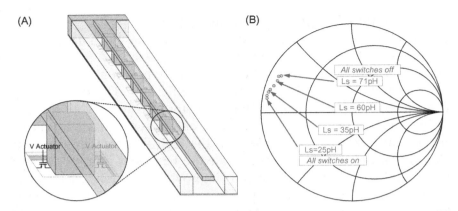

Figure 1.15 (A) Tunable transmission line stub with switches placed along the line to change stub inductance and (B) measurement results showing 3:1 inductance tuning ratio [8].

to change the effective value of the stub inductance, depending on the requirement. Switch transistors are deployed as shown in Figure 1.15A; these short the transmission line to the local ground at various points, leading to a change in inductance as depicted in the measurement results in Figure 1.15B.

1.1.5 Data Conversion and Healing Algorithm

To be able to interface the analog sensors with the on-chip digital optimi-zation component, ADCs must be implemented on-chip. These ADCs need to be low-power designs and occupy a low area; however, they need to operate fast enough to not be the bottleneck in the total healing time. In this design, 8-bit successive approximation-based ADCs were imple-mented that read the analog sensor outputs and convey the digitized serial data to the healing algorithm. A fully synchronous logic was selected and the functionality was verified up to 2.5 Msps (limited by the clock speed due to off-chip PCB clock routing and buffers). The average differential non-linearity (DNL) was −0.04 LSB, with a worst case of −0.605 LSB. This ensures that the ADC is monotonic, which ensures proper convergence of the healing algorithm.

The healing algorithm itself was coded in VHDL and synthesized using standard cells available for this technology. For scalable design, the algorithm should be implemented in a modular fashion with vari-ous components for sensor read, actuator write, and optimization. Depending on the nature of the search space itself, various optimiza-tion algorithms can be implemented, leading to faster healing times in

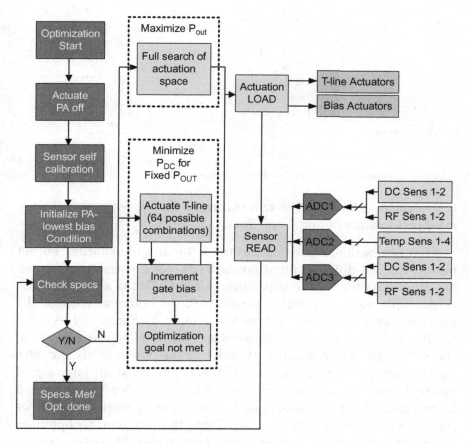

Figure 1.16 Flowchart showing two modes of self-healing [8].

real-life applications. In this implementation, the algorithm was based on a bulk search and had two modes of operation as shown in Figure 1.16: first is the mode that maximizes the output power and second is the mode that reaches a specified output power with minimum DC power consumption. The total number of states searched was approximately 262,000, with a worst-case healing time of approximately 0.8 seconds (limited by the digital clock speed).

1.1.6 Measurement Results

The PA was mounted on a PCB and probed at 28 GHz. The test setup is shown in Figure 1.17, with the output going to a power sensor through a calibrated mm-wave load tuner. In all the healing

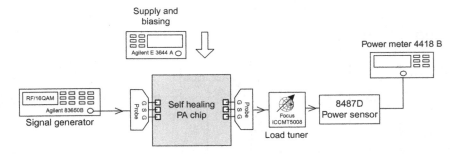

Figure 1.17 Measurement setup for the fully integrated self-healing PA [8].

measurements, the default state is chosen as the one with the best saturated performance during simulations.

First, the healing algorithm was run in mode 1 (maximizing output power) for two different input power levels—one at small signal and the other near the 1-dB compression point. Once the algorithm converged with maximum output power, with the optimum settings, the input power was varied to obtain the plots in Figure 1.18A. The results show an interesting trend—near the small-signal operation, the state that was optimized at small-signal power levels provides higher output power than the one that was optimized for large-signal power levels, whereas the reverse holds true as we move toward large-signal operation. This clearly shows one potential benefit of such a self-healing system and highlights the fact that no optimum actuation state exists that performs best across all output power levels. Measurements from 20 chips show similar improvements, as depicted in Figure 1.18B and C.

When the algorithm was run in mode 2, significant reductions in DC power consumption were observed across all power levels as depicted in the measurements from 20 chips in Figure 1.19. Because the state is not changing, the DC power without self-healing does not change appreciably over the entire range of power levels. On the contrary, with self-healing a dramatic improvement in DC power can be observed with up to 47% reduction at 12.5 dBm output power. Interestingly, the variation between the chips is also observed to decrease dramatically with self-healing. A 78% decrease in the standard deviation between chips is observed at the same output power of 12.5 dBm, demonstrating the advantages of self-healing in improving performance and yield.

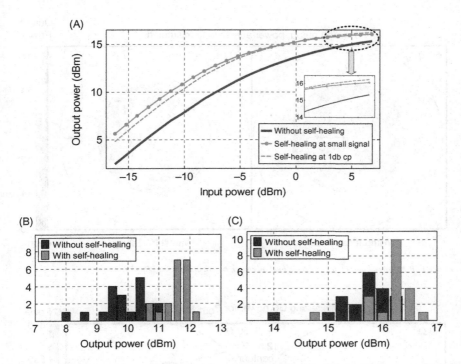

Figure 1.18 (A) Measured output power versus input power before and after self-healing. Output power healing for 20 chips at (B) small signal and (C) near 1-dB compression point [8].

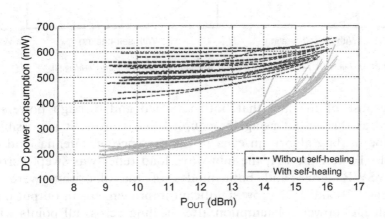

Figure 1.19 Measured DC power versus output power for 20 chips before and after healing [8].

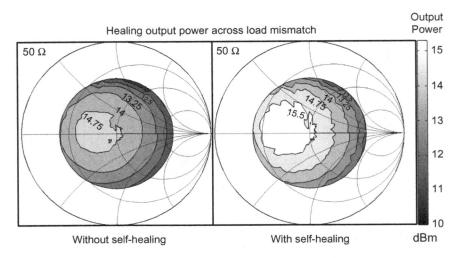

Figure 1.20 Measured output power across load mismatch before and after healing [8].

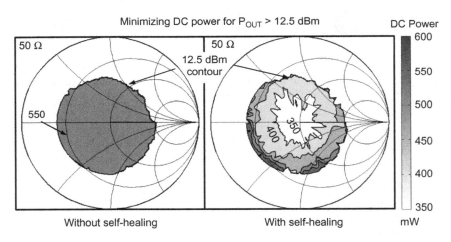

Figure 1.21 Measured DC power dissipation across load mismatch before and after healing [8].

Because of the fact that both reflected and through powers are sensed at the output using the coupled line sensors, it is possible to run the healing algorithm even in the presence of antenna load mismatch. To test the PA, the mm-wave load tuner was swept across a 4:1 VSWR circle and both modes of the algorithm were run. Figures 1.20 and 1.21 show significant improvements in output power and in DC power consumption after healing across all points within the swept range.

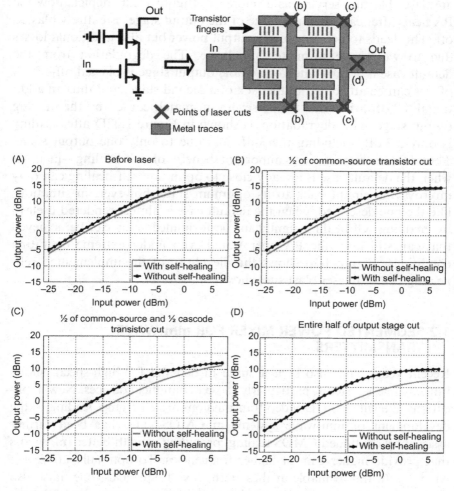

Figure 1.22 Schematic and layout location of sequential laser blasts as well as consequent improvements in output power once self-healing is enabled [8].

Finally, the self-healing PA was tested against partial or total transistor failure, which can be caused by aging and other transistor stress phenomena like voltage spikes. To demonstrate the flexibility of the self-healing system in dealing with variations it was not necessarily designed for, the output stage was repeatedly laser-blasted and at each step the healing algorithm for maximum output power was run. The output stage was chosen because of its high probability of failure due to the high-voltage swings it sustains. Figure 1.22A shows the PA without any laser-blasting serving as a reference. Figure 1.22C depicts how the PA heals after each step until one of the output stages is entirely blasted off. This leads to not only lower output power but also additional losses due to worst-case mismatch/imbalance. The degradation from the default case is 7.2 dB when one whole output stage is blasted off—3 dB of this can be attributed to the lack of a second stage, and thus an additional 4.2 dB arises due to the severe mismatch caused by the missing output stage. This degradation as shown in Figure 1.22D after healing is only 3.3 dB, including the 3-dB loss due to only one output stage. This demonstrates a very important benefit of self-healing—in cases when the default case is already close to optimum, self-healing can only improve the circuit back to its optimum state; however, in situations like the one described, when the default case is almost rendered useless, by enabling self-healing significant performance gains can be achieved that can still keep the PA operational even under extreme operating conditions. Figure 1.23 shows the die photo of the implemented self-healing PA with laser blast sites shown in the inset.

1.2 SEGMENTED POWER MIXER FOR mm-WAVE TRANSMITTERS

Recent years have also seen a surge in demand for high-speed, short-range wireless systems, particularly those in handheld devices. New applications like wireless HD video transmissions are emerging, leading to the development of mm-wave systems in CMOS. This has been further assisted by continued CMOS scaling, providing us with better transistors that provide higher gain at higher frequencies. To optimally utilize the wide bandwidth available at these mm-wave frequencies, one must also deploy complex nonconstant envelope modulation schemes that are typically more spectrally efficient than their constant envelope counterparts.

Linear amplification-based transmitters are well-equipped to handle nonconstant envelope modulation; however, they need to be operated significantly at back-off power levels, thereby sacrificing output power and efficiency. For example, the drain efficiency of a linear class-A PA decreases to approximately one-quarter of its peak value when operated at 6 dB below its saturated power level [10,11]. This decline in efficiency is of critical importance, particularly in handheld systems in which battery life is a premium. However, switching power amplifier-based transmitters can provide high output powers at high efficiencies, even at mm-wave

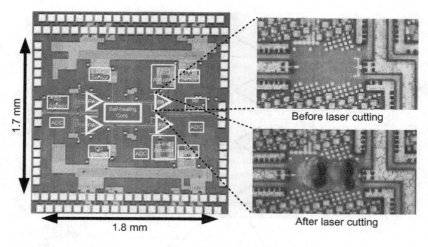

Figure 1.23 Die micrograph of the implemented system showing laser blast sites [8].

frequencies; however, their use is mostly limited to constant envelope modulation schemes, thus sacrificing spectral efficiency. At lower RF frequencies, switching PAs have been demonstrated to be suitable for nonconstant envelope modulation using supply modulators at mm-wave frequencies; due to the high data rates involved, this approach becomes impractical. In addition to this trade-off between linearity and power efficiency, other roadblocks exist in high-power, high-frequency CMOS designs primarily due to the low breakdown voltage of the transistors and the loss of on-chip power combining for fully integrated designs.

To address these trade-offs, a compact, energy-efficient, fully integrated mm-wave transmitter architecture is presented that is capable of generating complex nonconstant envelope modulations at high output power and at high data rates. The architecture is based on a polar scheme in which several segmented power mixers are power-combined on-chip using a dual-primary distributed active transformer (DAT) combiner [12]. Figure 1.24 depicts the overall chip architecture.

An input single-ended mm-wave LO signal is first converted to its differential form using an on-chip integrated balun, the output of which is then distributed to four differential power mixers using inverter-based driver stages. Baseband signals are applied to the IF ports of the power mixers in either pure digital or analog form, depending on the mode of operation. The differential output from all four power mixers is efficiently combined using the dual-primary DAT.

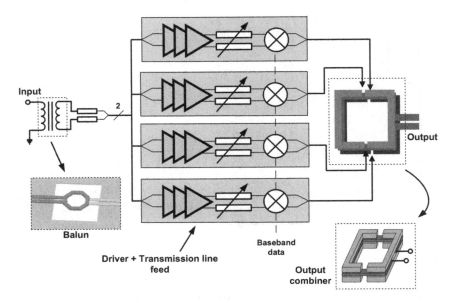

Figure 1.24 Overall architecture of power mixer transmitter.

Each power mixer is subdivided into eight smaller segments, with each segment operating at its full power, thereby maximizing its efficiency. Depending on the output power requirement, different segments can be turned ON or OFF, thereby also improving drain efficiency at back-off [13]. Direct amplitude modulation is also enabled through the same path, whereas phase modulation is provided through the mm-wave LO path. Figure 1.25 shows the example generation of a 16-QAM constellation using this architecture.

1.2.1 Key Building Blocks

The segmented power mixer was implemented in the 32-nm SOI CMOS process with 11 copper metal layers and 1 aluminum layer. The power mixers themselves are implemented as switched transconductor-based Gilbert cell stages (Figure 1.26A) with the bottom thin oxide transistors driven by the mm-wave LO and the thick-oxide upper quad driven by the baseband amplitude signals. As discussed previously, digital power control/modulation is enabled by dividing each power mixer into eight smaller segments, with seven of them being controlled digitally and the eighth segment being driven in an analog fashion, maintaining power continuity.

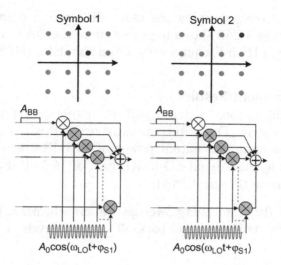

Figure 1.25 Example generation of two symbols in a 16-QAM constellation with phase modulation through mm-wave LO and amplitude modulation through digital baseband paths.

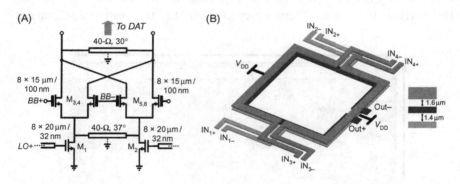

Figure 1.26 (A) Power mixer schematic and (B) output power combining structure [12].

Output power from the four differential power mixers are efficiently combined on-chip using the dual-primary DAT. The DAT is an efficient technique for combining output power from differential stages due to the formation of virtual shorts along the length of the transformer enabling convenient supply routing [14]. At mm-wave frequencies, its application has mostly been limited to output power combining from two differential stages primarily due to complexities involved in the input distribution network. In this design, the secondary is sandwiched between two primaries and each primary is driven by two differential power mixers. This enables a four-to-one differential power combining while simultaneously keeping the input distribution complexity at a minimum. Figure 1.26B shows the structure of the DAT combiner that was simulated in High Frequency Structure Simulator (HFSS).

1.2.2 Measurement Results

Figure 1.27 shows the die photo and the output power measurement setup for the system. The chip is mounted on PCB with input and output probed. A saturated output power of 19.1 dBm was achieved at 51 GHz, with a small-signal LO-to-RF gain of 16.2 dB that reduces to 7 dB in saturation (Figure 1.28A).

Because of the segmented power generation scheme, it is possible to improve drain efficiency at back-off power levels as shown in Figure 1.28B.

The power mixer was tested against various modulation schemes starting with constant envelope modulations (m-PSK). An arbitrary waveform generator was used to generate m-PSK modulations on an IF carrier that was then up-converted to the mm-wave carrier

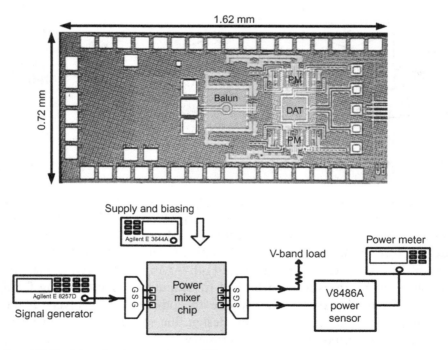

Figure 1.27 Die photograph and measurement setup [12].

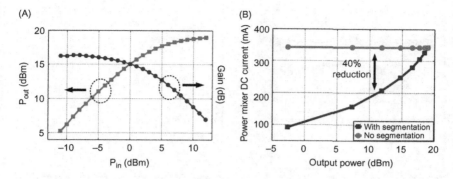

Figure 1.28 (A) Output power and gain versus input power and (B) DC power savings due to segmentation [12].

frequency using an external wide-IF mixer. The output of the chip was down-converted using an identical mixer and I-Q demodulation was performed in Matlab. Figure 1.29 shows measured eye diagram for a 4-Gbps BPSK modulation and recovered constellation for a 4-Gbps QPSK modulation.

The segmented power generation scheme also enables direct digital amplitude modulation, as shown in Figure 1.30. Binary ASK modulations were demonstrated at 1 Gbps (limited by the test setup) using this scheme. Depending on the number of segmentation levels, m-ASK modulations are also possible using this architecture, as also depicted in Figure 1.30.

By combining phase modulation through the mm-wave LO and amplitude modulations through the segmentation bits, it is possible to generate arbitrary m-QAM signals as shown in the measured constellation in Figure 1.31, where three amplitude levels are selected using the segmentation settings and 12 phase symbols are generated using the arbitrary waveform generator.

Another problem typically associated with segmented power generation is segment reliability. When the power mixer is operating at close to its maximum output power when one or two segments are OFF, the output voltage swing is still relatively large. This can produce very high voltage across the terminals of the off segment transistors, leading to catastrophic transistor breakdown or long-term reliability issues. In the present design, simulations were performed to ensure reliable segment operation under worst-case operation. The chip was measured for more than 8 hours with alternative switching

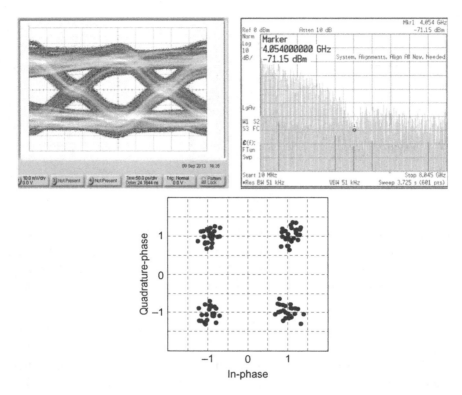

Figure 1.29 Measurement results from m-PSK modulations [12].

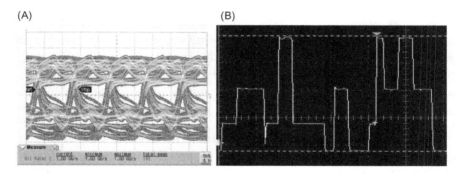

Figure 1.30 Measurement results from m-ASK modulations [12].

between all segments on and seven segments on, and while gradually increasing the stress on the eighth segment. No discernible degradation in output power was observed, even at a 30% higher supply voltage (Figure 1.32).

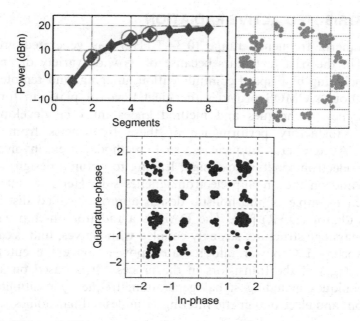

Figure 1.31 Generation of a 16-QAM constellation using this architecture [12].

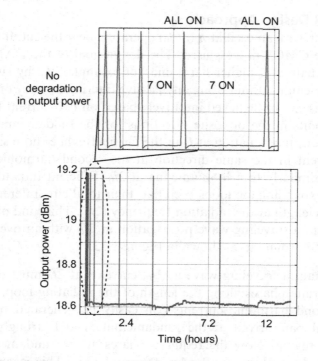

Figure 1.32 Segment reliability measurements at 30% higher V_{DD} [12].

1.3 DISTRIBUTED ACTIVE RADIATION

The terahertz frequency range (0.3–3 THz) has generated growing interest in the past few years because of its wide variety of applications, including high-speed communication, imaging, and remote sensing. Although CMOS provides the ideal low-cost platform for such systems, new techniques and methodologies must be developed to achieve satisfactory performance at these frequencies from MOS devices. At terahertz frequencies, such methods often involve true circuits–electromagnetics co-design, leading to optimal designs on the circuits and on the on-chip electromagnetics side. Here, we introduce one such example of the holistic design approach called distributed active radiator (DAR) [15]. The DAR is a radiator on-chip with an electromagnetic structure, strongly coupled with actives, that is capable of converting DC power into radiated power above the cutoff frequency (f_{max}) of the transistors in the process. It is based on several key techniques including signal generation, frequency multiplication, radiation, and electromagnetic filtering of undesired harmonics.

1.3.1 DAR Design Approach

Let us begin with an oscillator (at f_0) operating near the cutoff frequency f_{max} of the CMOS process used. The design goal of the DAR is to be able to radiate at a higher harmonic, for example, $2f_0$, by suppressing the fundamental without using any passive filtering or lossy on-chip elements. This can be achieved intuitively based on two current loops that carry currents in the opposite directions for the fundamental (thereby canceling out the radiation at f_0 in the far field) while those same loops carry current in the same direction at the second harmonic, thereby radiating similarly to a loop antenna at $2f_0$. Efficient radiators can be made by expanding the loops such that their total circumference equals a full wavelength at the radiation frequency ($2f_0$). This kind of loop can then sustain a traveling wave propagation at $2f_0$ with an overall phase shift of 360° around it, as shown in Figure 1.33.

To sustain a traveling wave at $2f_0$, currents at $2f_0$ must be injected at appropriate phases along the length of the radiating loop. Common mode second harmonic currents can easily be generated by using a differential pair driven at the fundamental f_0 and strongly into the nonlinear region. Now the required phases at the fundamental for a four-point driven loop are 0°, 45°, 90°, and 135°. This is conceptually

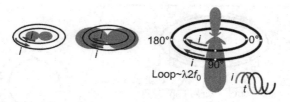

Figure 1.33 Canceling fundamental radiation to radiate only at the second harmonic using spatial filtering and a traveling wave radiator [15].

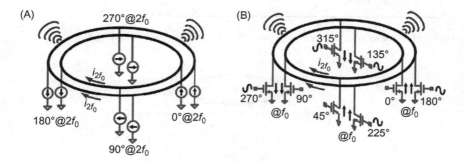

Figure 1.34 (A) Driving both loops at $2f_0$ in the proper phase progression and (B) pseudo-differential pairs driven at first harmonic to create the required currents at $2f_0$ [15].

illustrated in Figure 1.34, showing traveling waves at $2f_0$ in both loops propagating and radiating in phase.

However, the second harmonic currents are common mode and a return current path must be provided from source to sink. A simple ground plane can provide such a return path; however, then the loops essentially act as transmission lines and thus will be poor radiators. In this design, the return path is provided by pushing the return current away from the radiator with a ground aperture opening, as shown in Figure 1.35, where the radius of the cutout induces nearly a 90° phase difference between the forward radiating current and the return ground current such that the radiated fields due to these two currents add their power in space.

To generate a differential input swing at the gates of the transistors driving the loops, we cross-couple the transistors into a Mobius strip. This generates a total loop phase of 180° for the fundamental frequency, essentially acting as a differential transmission line for f_0 and all other odd harmonics. This is conceptually shown in Figure 1.36, where the $2f_0$ radiates primarily through the back side through the

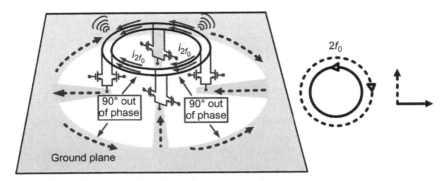

Figure 1.35 Second harmonic return current at nearly 90° out of phase with forward current [15].

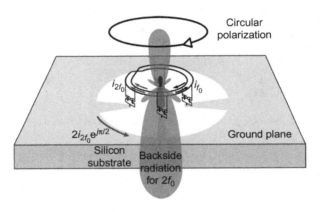

Figure 1.36 DAR showing Mobius strip achieving signal generation from DC, frequency multiplication based on device nonlinearity, filtering of the undesired odd harmonics, and radiation at the second harmonic [15].

high-dielectric constant Si substrate. It is evident that the radiated field is circularly polarized due to the traveling wave nature of the signal generation.

1.3.2 Architecture

To build large-scale arrays of such radiators to maximize the effective isotropic radiated power (EIRP) and to achieve beam steering for phased array applications, these DARs need to be mutually locked. Here, we demonstrate one such locking technique using transmission line networks and provide intuitive explanations for the same. If we place one coupling network between only two cross-coupled pairs, then it will ensure that they are driven in phase; however, the two DARs can still operate in opposite directions. By placing a second

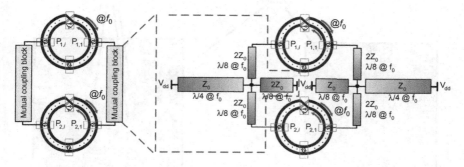

Figure 1.37 Locking using transmission line networks to synchronize two DARs [15].

coupling network as shown in Figure 1.37, it is now possible to ensure locking between the two DARs. The network needs to satisfy some properties to lock the DARs properly and at the fundamental.

Because the radiators need to be placed some distance apart for efficient radiation and to reduce mutual coupling, the locking networks will be essentially distributed in nature. Additionally, the network should not load the two DARs at both the fundamental and at the second harmonic. This can be ensured by using the network similar to that shown in Figure 1.37. This concept can also be extended to 2×2 DAR arrays.

1.3.3 Measurement Results

The 2×1 as well as 2×2 DARs based on transmission line locking were designed in the 45-nm SOI CMOS process with an estimated f_{max} of approximately 200 GHz. Fundamental frequency of oscillation was designed to be 150 GHz, which implies radiation at 300 GHz, beyond the f_{max} of the process. Figure 1.38 shows die photographs of the implemented chips.

Figure 1.39 shows the measurement setup for absolute power measurements. The chip was mounted on a PCB with a cutout to enable radiation from the back side. No external Si lenses or postprocessing was used.

The equivalent setup for measurement of radiation frequency is shown in Figure 1.40, where a harmonic mixer is used to down-convert the radiated signal.

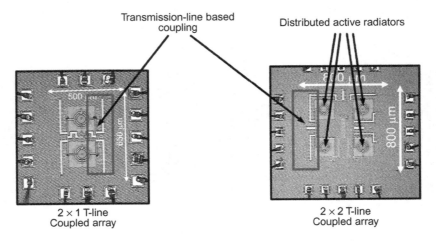

Figure 1.38 Die photos from 2 × 1 and 2 × 2 DAR arrays with transmission line-based synchronization [15].

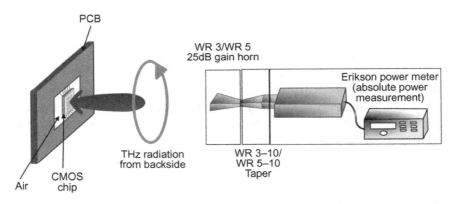

Figure 1.39 Calibrated measurement setup used to characterize radiated power.

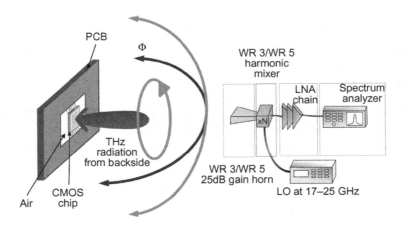

Figure 1.40 Measurement setup for measuring radiation frequency and radiation patterns.

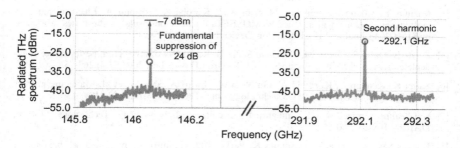

Figure 1.41 Calibrated spectra showing radiated second harmonic as well as fundamental suppression from the 2 × 2 array [15].

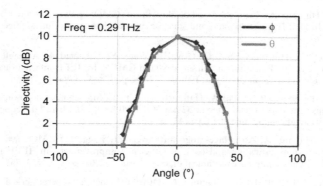

Figure 1.42 Measured radiation pattern in two orthogonal planes.

The circularly polarized signal radiated from the back side is detected at 299 GHz. The bore-sight EIRP was measured to be −13 dBm from the 2 × 1 array and −1 dBm for the 2 × 2 array. A fundamental suppression of 24 dB is observed because of the quasi-optical filtering effect of the DAR. Figures 1.41 and 1.42 show measurement results of the output spectrum as well as radiation patterns at 300 GHz.

REFERENCES

[1] Hajimiri A. mm-Wave silicon ICs: an opportunity for holistic design. In: IEEE radio frequency integrative circuits symposium; June 2008. p. 357−60.

[2] Hajimiri A. Holistic design in mm-wave silicon ICs. IEICE Trans Electron 2008;E91-C (6):817−28.

[3] Hoyt J, Nayfeh H, Eguchi S, Aberg I, Xia G, Drake T. et al. Strained silicon MOSFET technology. In: IEDM technical digest; December 2002. p. 23−6.

[4] Mizuno T, Takagi S, Sugiyama N, Satake H, Kurobe A, Toriumi A. Electron and hole mobility enhancement in strained-Si MOSFETs on SiGe-on-insulator substrates fabricated by SIMOX technology. IEEE Electron Device Lett 2000;21(5):230–2.

[5] Doyle B, Datta S, Doczy M, Hareland S, Jin B, Kavalieros J, et al. High performance fully-depleted tri-gate CMOS transistors. IEEE Electron Device Lett 2003;24(4):263–5.

[6] Kuhn K, Kenyon C, Kornfeld A, Liu M, Maheshwari A, Shih W, et al. Managing process variation in Intel's 45 nm CMOS technology. Intel Tech J 2008;12(2):93–109.

[7] Self-healing mixed-signal integrated circuits. Microsystems Technology Office, DARPA-BAA-08-40.

[8] Bowers SM, Sengupta K, Dasgupta K, Parker BD, Hajimiri A. Integrated self-healing for mm-wave power amplifiers. IEEE Trans Microw Theory Tech 2013;61(3):1301–15.

[9] Sengupta K, Dasgupta K, Bowers SM, HajimiriA. On-chip sensing and actuation methods for integrated self-healing mm-wave CMOS power amplifiers. In: IEEE MTT-S international microwave symposium digest; June 2012. p. 1–3.

[10] Kang J, Hajimiri A, Kim B. A single-chip linear CMOS power amplifier for 2.4 GHz WLAN. In: IEEE international solid-state circuits conferences technical digest; February 2006. p. 761–9.

[11] Chowdhury D, Hull C, Degani O, Goyal P, Wang Y, Niknejad A. A single-chip highly linear 2.4 GHz 30 dBm power amplifier in 90 nm CMOS. In: IEEE international solid-state circuits conferences technical digest; February 2009. p. 378–9, 379a.

[12] Dasgupta K, Sengupta K, Pai A, Hajimiri A., A 19.1 dBm segmented power-mixer based multi-Gbps mm-Wave transmitter in 32 nm SOI CMOS. In: IEEE radio frequency integrative circuits symposium; June 2014. p. 343–6.

[13] Kousai S, Hajimiri A. An octave-range watt-level fully integrated CMOS switching power mixer array for linearization and back-off efficiency improvement. In: IEEE international solid-state circuits conferences technical digest; February 2009. p. 376–7, 377a.

[14] Aoki I, Kee S, Rutledge D, Hajimiri A. Distributed active transformer—a new power-combining and impedance-transformation technique. IEEE Trans Microw Theory Tech 2002;50(1):316–31.

[15] Sengupta K, Hajimiri A. Distributed active radiation for THz signal generation. In: IEEE international solid-state circuits conferences technical digest; February 2011. p. 288–9.

CHAPTER 2

Cartesian Feedback with Digital Enhancement for CMOS RF Transmitter

Nathalie Deltimple[1], Bertrand Le Gal[1], Chiheb Rebai[2], Alexis Aulery[1], Nicolas Delaunay[1,3], Dominique Dallet[1], Didier Belot[3,4] and Eric Kerhervé[1]

[1]University of Bordeaux, IMS Laboratory, CNRS UMR 5218, Bordeaux INP France, [2]GRES'COM Lab, SUP'COM, University of Carthage, Tunisia, [3]STMicroelectronics, Rue Jean Monnet, Crolles, [4]CEA-LETI, Rue de Martyrs, Grenoble

2.1 INTRODUCTION

Power efficiency is one of the most important parameters leading to green electronics. However, high data rate signals require linearity to overcome amplitude and phase distortions. Knowing that power amplifiers consume a lot of power, one way to achieve this is to develop integrated linearization techniques for CMOS technology. In the literature, there are several ways to linearize power amplifiers [1−11]. For feedback, feedforward, or predistortions, the goal is to fix nonlinearities. Other methodologies are used to avoid these distortions, such as a constant envelope signal applied on the power amplifier as in EER (envelope elimination and restoration) architecture, LINC (linear amplification with nonlinear components), or CALLUM (constant amplitude locked loop universal modulator). Among all these possibilities, we chose to study a fully integrated Cartesian feedback (CFB) technique in 65-nm CMOS technology. This technique has the main advantages of not requiring a nonlinear model and being insensitive to process (or temperature, aging, and so on) variations.

In this chapter, we describe the CFB technique and introduce the original architecture based on a mixed signal solution, an alternative to the full-analog solution presented previously [4]. Simulations and measurements are reported to demonstrate the proposed solution efficiency.

Linearization and Efficiency Enhancement Techniques for Silicon Power Amplifiers.
DOI: http://dx.doi.org/10.1016/B978-0-12-418678-1.00002-7

2.2 CFB LOOP

The idea of using CFB to linearize power amplifiers has been discussed as early as the 1970s. The linearization technique using a CFB was studied by Cox in 1975 [12], and then by Petrovic in 1979 [13]. The technique is called Cartesian feedback because the feedback is based on the Cartesian coordinates of the baseband symbols, I and Q, as opposed to the polar coordinates. A typical CFB architecture is illustrated in Figure 2.1. The top of the architecture is the Zero-IF transmitter chain, which is called the direct chain. The loop is then closed at baseband, rather than at the carrier frequency, and down-converters, in association with filters used to remove the intermodulation products (at $2\omega_{LO}$) generated by the down conversion, are needed to realize this operation. Hence, the linearization is performed only for a narrow band signal instead of from DC to the carrier.

The I and Q paths are treated with two identical loops, and this architecture is quite simple to use.

Nevertheless, Dawson and Lee have shown that the delay through the power amplifier, the phase shifts of the RF carrier attributable to the reactive load of the antenna, and the mismatched interconnect

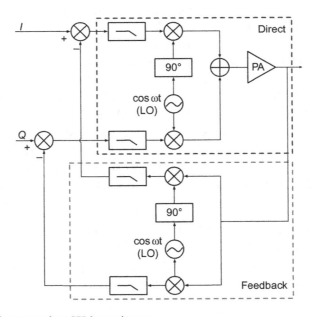

Figure 2.1 PA linearization by a CFB loop architecture.

lengths between the local oscillator (LO) source and the two mixers all manifest as an effective nonzero phase shift and the signal provided by the LO $cos\,\omega t$ becomes $cos(\omega t + \varphi)$ [4]. This phase misalignment can lead to instability, and stability is the major issue of this architecture because of the feedback path. Hence, this misalignment needs to be compensated by adding a phase alignment regulator in the CFB system, as shown in Figure 2.2.

Another important advantage of the CFB technique is that it can linearize the power amplifier, whatever the process or aging. Hence, the value of the phase in the "phase compensation" function has to be controllable.

Based on the architecture developed by Dawson, we studied digital phase compensation architecture [14]. Our goal was to obtain a multistandard linearization technique that provides flexibility in the control of phase compensation and stability. This led to the architecture depicted in Figure 2.3.

Compared with the original architecture depicted in Figure 2.1, we realized controllable phase compensation operation and subtraction in the digital domain. This solution that performs compensation in the

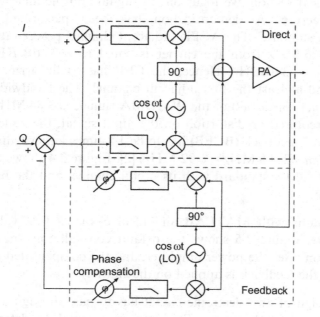

Figure 2.2 CFB loop architecture with phase compensation.

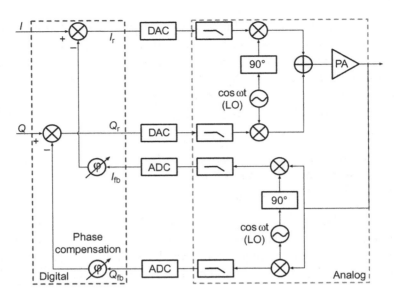

Figure 2.3 Mixed analog/digital CFB loop architecture.

digital domain requires analog-to-digital converters and digital-to-analog converters.

In the next section, we focus on the digital and the analog parts of the architecture. A demonstration has been performed for the W-CDMA standard. The ACPR (adjacent channel power ratio) specification at 5 MHz from the carrier is equal to −33 dBc/Hz and at 10 MHz is −43 dBc/Hz. Hence, the CFB has to linearize the main channel and at least the first adjacent channel. The bandwidth of the main channel (*Bchannel*) of the W-CDMA signals is 3.84 MHz and, to cover the required predistortion on the input signal, the bandwidth of the Cartesian feedback (BCFB) has to be 1.8-times larger (which represents *Bchannel*/2 + 5 MHz = 6.92 MHz) [2]. Figure 2.4 shows the signal of the W-CDMA standard with its useful channel and the bandwidth of the CFB.

Simulations with MATLAB and ideal elements first validate the architecture. Figure 2.5 shows the default constellation, the distorted constellation after the power amplifier, and the compensated constellation when the feedback is applied on the emitter.

The design of any electronic circuit has to pass through a first step called system-level behavior. This step is essential to determine the

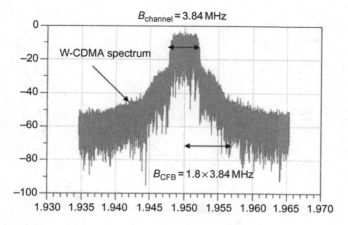

Figure 2.4 Bandwidth of the Cartesian feedback.

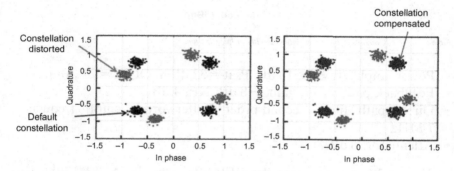

Figure 2.5 Default, distorted, and compensated constellation.

characteristics of all the building blocks of the circuit in the transmitter, receiver, or transceiver. ADS software from Agilent was used to acquire a first idea of the components' properties in this study of CFB.

This architecture is composed of low-pass filters, up-conversion mixers, a power amplifier, a coupler for the direct path, and an attenuator, a splitter, down-conversion mixers, and low-pass filters for the feedback path. An ideal phase shifter and a subtractor correct the system. First, a literature study is performed to determine typical parameters of each block in the technology of integration, i.e., CMOS 65-nm technology from STMicroelectronics.

- Direct path mixer: gain, 14 dB; NF, 6.4 dB
- Direct path filter (second-order Butterworth), cutoff frequency: 7 MHz

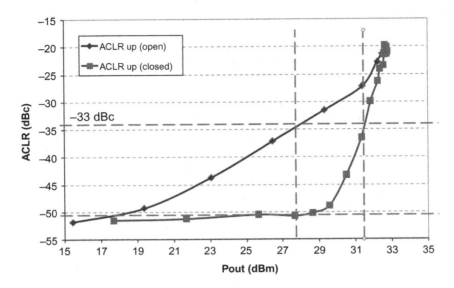

Figure 2.6 ACLR versus output power with and without linearization.

- Power amplifier: gain, 15 dB; Psat = 32 dBm; OCP1 = 27 dBm
- Feedback path mixer: gain, −8 dB; NF, 4 dB
- Direct path filter (second-order Butterworth), cutoff frequency: 7 MHz
- Attenuator: 6 dB

Figure 2.6 demonstrates the CFB effect on the ACLR1 (the first adjacent channel). At 5 MHz from the carrier, the ACLR specification is −33 dBc. Without CFB, this value is reached for output power of 28 dBm. When the loop is closed, ACLR1 is equal to −51 dBc (18 dB of improvement) and the −33 dBc value is reached for an output power of 31.5 dBm (3.5 dB of improvement).

System simulations realized with MATLAB and ADS validate the proposed CFB architecture. In the next section, both digital and analog parts are described and integrated in 65-nm CMOS technology from STMicroelectronics.

2.3 CFB DIGITAL PART IMPLEMENTATION

The main task of the digital part is to realize the phase compensation and the subtraction of the feedback on direct I and Q paths. The first step is to evaluate the θ angle of the phase correction. This angle is calculated by comparing forward and feedback paths phases. Therefore, the digital CFB architecture is decomposed in three main parts:

- Phase estimation
- Vector rotation
- Subtraction

Two architectures are evaluated for the circular transform implementation:

- The first architecture is based on lookup tables (LUTs) and multipliers. The main advantage of this approach concerns the processing latency that is quite low. To avoid considerable delay into the loop, this parameter should be maintained as low as possible. However, this approach also has the drawback of its expensive cost in terms of silicon area.
- The second architecture uses Coordinate Rotation Digital Computer (CORDIC) operators [15], which consume less area than the multipliers when the data path exceeds 10 bits. To achieve high working frequency, pipelined architectures of the CORDIC have been considered. Nevertheless, it introduces higher latency in the loop and a trade-off between area occupation, latency, and throughput has to be defined. Fine-tuning of the implementation constraints leads to an optimal solution.

Delay in the loop is limited by W-CDMA data (T_{chip}) for stability considerations [4]. The input and output of the digital stage are synchronized with the DAC and ADC frequency specifications, and the sampling frequency is set to 242 MHz.

2.3.1 Phase Estimation

Regarding phase estimation, it is very important to notice that phase subtraction must be performed "modulo 2π" to keep the same range of variation of the angle applied to the next block (vector rotation). It implies that the phase estimation block is divided into two subfunctional units: "atan" unit for the direct and feedback phase estimations and "modulo-2π" unit to limit the range.

2.3.1.1 *"atan"-Based Architecture*

Different implementations of the digital "atan" function are proposed in the literature [16]. One of them that is the most trivial uses a LUT [17]. This solution seems to be significant and very appealing in silicon area

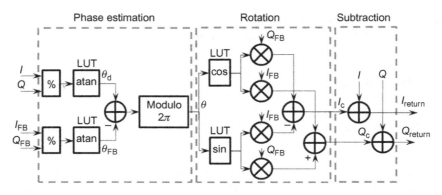

Figure 2.7 LUT-based solution.

consumption compared with other alternative implementations such as the CORDIC algorithm. A comparative study regarding implementation of these two solutions is presented, as well as output performances.

2.3.1.1.1 LUT-Based Architecture

LUT-based solution, illustrated in Figure 2.7, consists of first making the division of "Q" by "I", and then going through an interpolation table (LUT) where the required values of the "atan" function are stored. LUT length is function of the phase estimation precision. Precision of 1° (required in the current study) leads to 1440 bits of ROM. Because of the symmetry properties of "atan" function, the LUT size can be limited to half [8].

Several architectures were proposed in the literature to implement the division, but the so-called Restoring Division Algorithm [18] was chosen because it enables higher working frequencies. This architecture requires both operands to be positive and for the numerator to be greater than or equal to the denominator. Additional developments can process negative operands. Division by "0" results in saturation at the maximum value.

2.3.1.1.2 CORDIC-Based Architecture

Another alternative to implement the "atan" function is to use the iterative CORDIC algorithm [19]. The CORDIC-based solution does not require division, as shown in Figure 2.8. It takes the two coordinates of the vector as input and directly provides its phase [19]. This algorithm was designed to use only shift registers and adder/subtractor hardware resources, as presented in Figure 2.9. It was subsequently

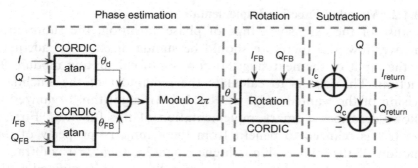

Figure 2.8 CORDIC-based solution.

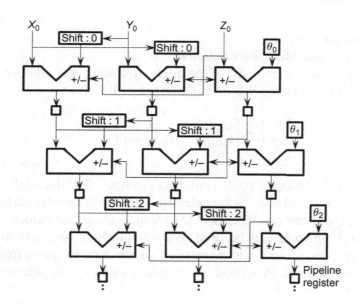

Figure 2.9 CORDIC example implementation.

improved to calculate trigonometric functions (exp, cos, sin, atan, and others), and this is accomplished by wisely configuring its input variables.

CORDIC implementation can be performed in several ways [20]. A trade-off between throughput and design area is required. Its small computation complexity and modularity provide considerable flexibility during the implementation process, and that is why partially pipelined architecture was chosen to meet frequency and latency requirements.

2.3.1.2 "Modulo" Function Implementation

Because of the extensive range of phase variation, the subtraction can overflow and therefore should be standardized to be adequate for the input of the next stage. It consists of calculating modulo 2π function. The idea is to calculate the remainder of the Euclidean division of the wanted angle by $K \times 2\pi$. An algorithm organized as follows describes a smart implementation of this function. First, a sign test is realized to benefit from the symmetry property of this function. If the angle value is higher than 2π, then 2π is subtracted from it and the test continues; otherwise, this value is retained as the output result.

2.3.2 Vector Rotation

2.3.2.1 LUT and Multipliers Solution

This first solution consists of precisely following the mathematical function of the rotation vector as described in Eq. (2.1).

$$\begin{bmatrix} I' \\ Q' \end{bmatrix} = \begin{bmatrix} \cos(\theta) & -\sin(\theta) \\ \sin(\theta) & \cos(\theta) \end{bmatrix} \begin{bmatrix} I_{FB} \\ Q_{FB} \end{bmatrix} \qquad (2.1)$$

Figure 2.7 shows an implementation example. Because of the phase cosine and sine values, the complex multiplication can be performed. To compute those coefficients, LUTs that contain the values of sine and cosine function can be used. An optimization step uses a single LUT (e.g., cosine) and trigonometric relationships to move from one function to another. A second optimization step exploits quarter wave symmetry.

2.3.2.2 CORDIC-Based Solution

As shown in Figure 2.8, the CORDIC algorithm can perform a vector rotation without multiplier resources allocation using simple initiation of entries of the vector on which we rotate, namely IFB/QFB, and the performance angle. Indeed, this architecture was chosen to realize vector rotation function.

2.3.3 Subtraction

Subtraction is simple enough to be implemented in digital domain. It is necessary to calculate the complement of the second operand and then use an adder.

2.3.4 Improvement of CFB CORDIC-Based Architecture

Existing *atan* CORDICs architecture allows calculation of angles using Cartesian coordinates of vectors. The CFB algorithm needs two *atan* blocks, followed by a subtraction of the estimated angles. A new architecture has been formed merging two *atan* CORDICs to directly compute the difference of angles [21]. Resulting architecture is scalable, like all other CORDIC architecture.

2.3.4.1 Modification of the CFB Algorithm

The *atan* CORDIC computes angles by iterative pseudo-rotations. For each iteration, the actual angle is added or subtracted by the angle of the pseudo-rotation, depending on the direction. Angles of pseudo-rotation are constant and already known by the system. If θ_{fw} is the estimated angle of command and n is the number of iterations, then it will be expressed by:

$$\theta_{fw} = \pm\,\theta_0 \pm \theta_1 \pm \cdots \pm \theta_{n-1} \tag{2.2}$$

It appears that the final result can be stored or generated only by knowing the consecutive directions of pseudo-rotation. On the initial design, the two angles θ_{fw} and θ_{fb} are estimated by two identical *atan* CORDIC blocks and then are subtracted. Therefore, results of subtraction can be expressed by:

$$\theta_{fw} - \theta_{fb} = f_0\theta_0 + f_1\theta_1 + \cdots + f_{n-1}\theta_{n-1} \tag{2.3}$$

This means that the result of subtraction can also be stored or generated by knowing the consecutive directions of rotation of the *atan* CORDICs. The key to reducing redundant calculations is to immediately compute subtraction, depending on direction of rotation of the two *atan* CORDICs.

$$f_i = \begin{cases} 0 & \text{if } f_w \text{ and } f_b \text{ same sign} \\ 2 & \text{if } f_w > 0 \text{ and } f_b < 0 \\ -2 & \text{if } f_w < 0 \text{ and } f_b > 0 \end{cases} \tag{2.4}$$

2.3.4.2 New Architecture

All these considerations lead to a new architecture (Figure 2.10) based on the initial one shown in Figure 2.9. Each horizontal stage is composed of five operators and corresponds to an iteration of the algorithm. As shown in Figure 2.10, the result of the merged

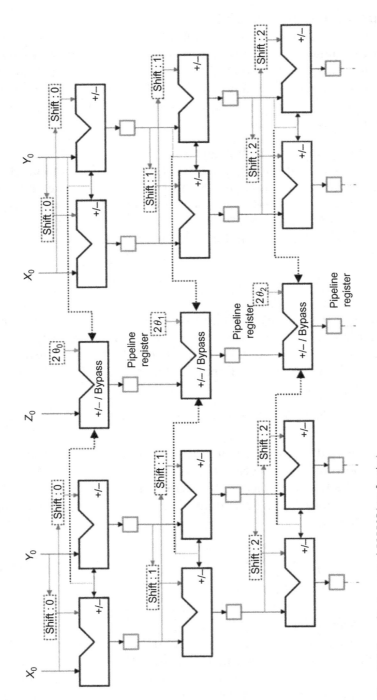

Figure 2.10 New architecture: merged CORDIC (atan function).

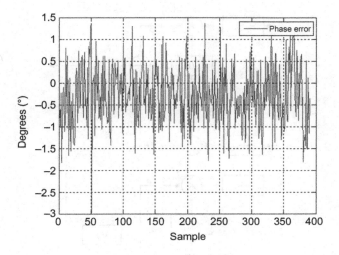

Figure 2.11 Phase error result of merged CFB.

CORDIC architecture is a new design with lower complexity than the previous one in terms of operators. This design is scalable in the same way as other CORDIC architectures.

With this new design, functionality and precision are preserved. Moreover, the critical path is unchanged, leading to the same working frequency. To keep the same latency, we just have to repeat the same pipelining model used previously [22].

Simulations have been performed to confirm that functionality and precision are preserved. Results of Modelsim simulations are depicted in Figure 2.11. Note that 1° precision is not absolute precision and it can never be achieved because of "quantum error" of digital systems. However, precision is reached in terms of average and standard deviation.

2.4 ANALOG PART IMPLEMENTATION

Analog relies on the design of mixers, filters, the attenuator, and the power amplifier in the direct path and in the feedback path, as seen in Figure 2.12 showing the block diagram point of view and as shown in Figure 2.13 in the die photograph. The technology used is CMOS 65 nm from STMicroelectronics.

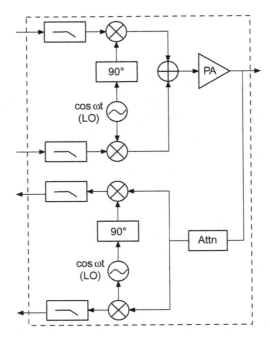

Figure 2.12 Analog part of the loop.

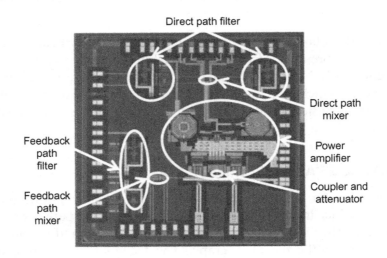

Figure 2.13 Die photograph of the analog part (2 mm × 2 mm with pads).

The direct path on each I and Q branch is composed of:

- A Gm-C filter used to select the useful channel and filter the spurious generated by the digital-to-analog converter. The filter comprises three stages. The input stage is a first-order passive filter without tuning frequency capability. This filter has a 3-dB cutoff frequency equal to 10 MHz. This stage is used to match the impedance between the block placed before the filter (the converter or the passive mixer) and the PMA (pulse with modulation amplifier). The second stage (PMA) is an amplifier designed with linear transconductance to increase the in-band linearity. The variation gain depends on the desired standard (8, 18, or 28 dB). It is the most limited design because of the linearity and noise constraints (applied to the feedback path). Finally, the last stage is a first-order active filter with a 3-dB cutoff frequency equal to 8 MHz. This cutoff frequency can be tuned at $\pm 10\%$ to ensure the useful signal. The die area of the active filter is 360 μm $\times$ 360 μm.
- An active mixer based on a Gilbert cell mixer with a 16-dB gain for a maximum output power of 10 dBm and a 1-dB compression point of -17 dBm. The die area of the active mixer is 270 μm $\times$ 380 μm.
- A class-AB power amplifier based on a two-stage cascade and differential architectures. The power amplifier has been designed and measured by STEricsson. Measured power gain is 24 dB, OCP1 is 22.8 dBm, and saturated power is 26 dBm. The die area of the PA is 1.3 mm^2.

Regarding the feedback path, on each I and Q branch we found:

- A passive mixer based on a quadrature ring structure. The main advantages are the power consumption and the large bandwidth; 3 dB of attenuation is observed, which is not a problem in the feedback path.
- A variable attenuator from 2 to 16 dB of attenuation. The die area is 71 μm $\times$ 63 μm.
- A Gm-C filter. The same architecture is used in the direct path. The die area is 5.5 μm $\times$ 24 μm.

2.5 LINEARIZED TRANSMITTER RESULTS

To evaluate the mixed CFB loop, some co-simulations are performed between the analog part and the digital part with SystemVue Electronic

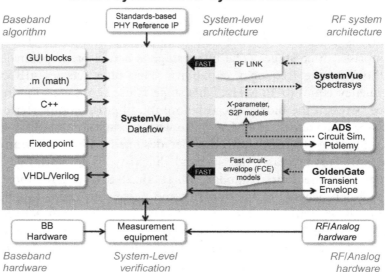

Figure 2.14 SystemVue RF System Architect combines baseband algorithms with RF system architectures for a more realistic view of the overall system performance [23].

System Level (ESL) design software from Agilent. It enables system architecture development because SystemVue also connects with circuit level design flows, such as VHDL code with ModelSim, ADS (Ptolemy, circuit simulation), or Fast Circuit Envelope models (from GoldenGate), or measurement results. The software configuration is shown in Figure 2.14.

Linearity improvement can be observed in Figure 2.15. In the open loop configuration, the $ACLR_1$ (at 5 MHz from carrier) limit of -33 dBc is reached for output power of 13 dBm. In the closed loop configuration, the limit of -33 dBc is reached for output power of 24 dBm. This improvement allows the ACLR specifications to be achieved. PA OCP1 is 22.8 dBm; at this output power, ACLR is enhanced from 41 dBc (open loop) to -9.5 dBc (closed loop). The maximum ACLR improvement is 34 dB.

2.6 POWER CONSUMPTION AND SIZE CONSIDERATIONS

Adding functions around the power amplifier is the main drawback of any linearization technique regarding the increase of power

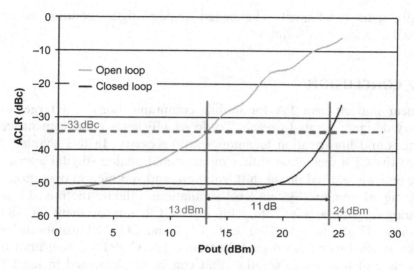

Figure 2.15 ACLR versus output power in open loop and closed loop configurations.

Table 2.1 Power Consumption of Each Element of the Architecture					
Direct path (mW)					
Gm-C Filters	**Active Mixers**	**Power Amplifier**		**DAC**	**Total**
20.5	67.75	@P_{out} = 22.8 dBm: 658.4		A: 26; D: 2.5	891.9
Feedback path (mW)					
Attenuator	**Passive Mixers**	**Gm-C Filters**	**ADC**	**Digital Functions**	**Total**
1.25	0	24.9	A: 90; D: 1	4.7	237.8

consumption and die area. Hence, for the mixed CFB technique, we had to evaluate power consumption of the feedback path and compare that with the direct path (Table 2.1).

The feedback path consumes 237.9 mW, whereas the direct path consumes 891.9 mW (for output power of 22.8 dBm). The power consumption of the mixed CFB is 1130 mW. The feedback path consumes 21% of the total power consumption. Note that without the feedback path, output PA linearity specifications are not satisfied. The digital part consumes only 4.7 mW. For comparison [24], the digital functions of the digital CFB architecture consume 33.3 mW.

The area of the analog part is 2.3 mm^2 (without pads). The area of the data converters is 1.55 mm^2 (0.476 mm^2 for the DAC and 0.3 mm^2 for the ADC). Approximately 1% of the total area is occupied by the digital part $(0.04 \, \text{mm}^2)$. The mixed analog–digital solution is well-suited as a low-cost solution.

2.7 CONCLUSION

Linear and efficient PA for mobile communications is a target that every designer wants to reach. To do so, efficiency enhancement techniques and linearization techniques are necessary. In this chapter, we used the CFB technique with a mixed signal analog–digital approach because we realized phase shift, rotation, and subtraction operations in the digital domain. System-level simulations validate the mixed signal analog–digital architecture with an ACLR improvement of 18 dB (from -33 dBc to -51 dBc). Both LUT and CORDIC solutions have been studied to realize digital functions. The CORDIC algorithm is a simple and low-power solution that can be implemented in the CFB technique. Digital, conversion, and analog parts are implemented in 65-nm CMOS technology from STMicroelectronics. The results obtained with system-level simulations based on SystemVue software and measurement results allow ACLR specifications for a maximum output power of 24 dBm at 1.95 GHz.

REFERENCES

[1] Yi J, Yang Y, Park M, Kang W, Kim B. Analog predistortion linearizer for high-power RF amplifiers. IEEE Trans Microw Theory Tech 2000;48(12):2709–13.

[2] Katz A. Linearization: reducing distortion in power amplifiers. IEEE Microw Mag 2001;2(4):37–49.

[3] Carrara F, Scuderi A, Palmisano G. Wide-bandwidth fully integrated Cartesian feedback transmitter. In: Proceedings of the IEEE 2003 CICC, custom integrated circuits conference. p. 451–54.

[4] Dawson JL, Lee TH. Automatic phase alignment for a fully integrated Cartesian feedback power amplifier system. IEEE, Solid-State Circuits 2003;38:2269–79.

[5] McCune E, Sander W. EDGE transmitter alternative using nonlinear polar modulation. In: Proceedings of the 2003 international symposium on circuits and systems (ISCAS'03); 2003. p. III-594–III-597.

[6] Boumaiza S, Jing L, Jaidane-Saidane M, Ghannouchi FM. Adaptive digital/RF predistortion using a nonuniform LUT indexing function with built-in dependence on the amplifier nonlinearity. IEEE Trans Microw Theory Tech 2004;52(12):2670–7.

[7] Sowlati T, Rozenblit D, Pullela R, Damgaard M, McCarthy E, Dongsoo K, et al. Quad-band GSM/GPRS/EDGE polar loop transmitter. IEEE J Solid-State Circuits 2004;39 (12):2179–89.

[8] Chung SW, Holloway JW, Dawson JL. Open-loop digital predistortion using Cartesian feedback for adaptive RF power amplifier linearization. IEEE 2007;1449–52.

[9] Presti CD, Carrara F, Scuderi A, Asbeck PM, Palmisano G. A 25 dBm digitally modulated CMOS power amplifier for WCDMA/EDGE/OFDM with adaptive digital predistortion and efficient power control. IEEE J Solid-State Circuits 2009;44(7):1883–96.

[10] Boo HH, Chung SW, Dawson JL. Digitally assisted feedforward compensation of Cartesian-feedback power-amplifier systems. IEEE Trans Circuits SystII Express Briefs 2011;58(8).

[11] Presti CD, Kimball D, Asbeck P. Closed-loop digital predistortion system with fast real-time adaptation applied to a handset WCDMA PA module. IEEE Trans Microw Theory Tech 2012;60(3, Part: 1):604–18.

[12] Cox D. Linear amplification by sampling techniques: a new application for delta coders. IEEE Trans Commun 1975;23(8):793–8.

[13] Petrovic V, Gosling W. Polar-loop transmitter. Electron Lett 1979;15(10):286–8.

[14] Delaunay N, Deltimple N, Kerhervé E, Belot D. A RF transmitter linearized using Cartesian feedback in CMOS 65 nm for UMTS Standard. In: IEEE 2011 topical conference on power amplifier for wireless and radio applications (PAWR-RWS'11), Phoenix, AZ, January; 2011. p. 16–20.

[15] Andraka R. A survey of CORDIC algorithms for FPGA based computers. In: Proceedings of the 1998 ACM/SIGDA sixth international symposium on field programmable gate arrays. ACM; 1998. p. 191–200.

[16] Altmeyer RC. Design implementation, and testing of a high performance ASIC for extracting the phase of a complex signal [Ph.D. dissertation]. California: Naval Postgraduate School Monterey; 2002.

[17] Schwarzbacher AT, Brasching A, Wahl TH, Foley JB. Optimisation and implementation of the arctan function for the power domain. Electronic circuits and systems conference, Bratislava, Slovakia; 1999.

[18] Oberman S, Flynn M. Division algorithms and implementations. IEEE Trans Comput 1997;46(8):833–54.

[19] Volder JE. The CORDIC trigonometric computing technique. IRE Trans, Electronic Comput 1959;EC-8:330–4.

[20] Weber J, Meaudre M. Le Langage VHDL. France: DUNOD; 1997.

[21] Aulery A, Dallet D, Le Gal B, Deltimple N, Belot D, Kerherve E. Study and analysis of a new implementation of a mixed-signal Cartesian feedback for a low power zero-IF WCDMA transmitter. IEEE International NEWCAS'13, Paris, France, June 16–19; 2013. p. 1–4.

[22] Sanaa W, Delaunay N, Le Gal B, Dallet D, Rebai C, Deltimple N, et al. Design of a mixed-signal Cartesian feedback loop for a low power zero-IF WCDMA transmitter. In: IEEE 2012 Second Latin American Symposium on Circuits and Systems (LASCAS'12).

[23] SystemVue Electronic System-Level (ESL) Design Software, < http://www.home.agilent.com >.

[24] Viteri A, Zjajo A, Hamoen T, van der Meijs N. Digital Cartesian feedback linearization of switched mode power amplifiers. 17th IEEE international conference on electronics, circuits, and systems (ICECS); 2010. p. 890–93.

CHAPTER 3

Transmitter Linearity and Energy Efficiency

Earl McCune
RF Communications Consulting, Santa Clara, CA

3.1 INTRODUCTION

One of the most well-known trade-offs in transmitter design is between linear performance of the power amplifier (PA) and energy efficiency of this amplifier [1]. Every marketing person I have encountered insists that both of these performance criteria are "first priority," meaning that their desire is to have engineers deliver perfect linearity at consistently perfect energy efficiency. Unfortunately, physics does not support this combination in the same circuit; you get to choose which one will perform maximally, and the other will degrade.

In this chapter we first carefully examine the PA development problem in modern transmitter design, noting how the signal modulations selected affect the difficulty or ease of transmitter design. The next step considers a "backwards" design approach, which, instead of starting with a linear circuit and then trying to improve its energy efficiency, starts with a maximally energy-efficient circuit and works to make it behave "linearly." Performance expectations from such a design are developed, and then data are reported to evaluate these predictions.

3.2 THE PA DESIGN PROBLEM

PAs are peak power limited circuits, meaning that any PA can provide only so much power into its load. Any particular PA must be capable of providing the maximum output power that the signal will ever require, called the peak envelope power (PEP). This output power is different from that used in the radio communications design process, where coverage and range are dependent on the signal average power. The ratio of PEP to the average power of any signal is called the peak-to-average power ratio (PAPR).

Linearization and Efficiency Enhancement Techniques for Silicon Power Amplifiers.
DOI: http://dx.doi.org/10.1016/B978-0-12-418678-1.00003-9

This leaves a dichotomy. The system design uses average power, but the PA design must be based on PEP [2]. This means that the PA design must be larger than that envisioned from the system design, where the scale factor is the linear equivalent of the PAPR (which is commonly expressed in decibels). How this scale factor works is illustrated in Figure 3.1. The different lines correspond to how close to the amplifier saturated output power the signal peaks are allowed to get, the output back-off (OBO). At 0 dB, the output signal peaks are allowed to be at the PA saturated power level. This situation still distorts the signal peaks, so that linearity is improved by limiting the signal peaks to approach only within 2 dB or 4 dB of the PA saturated power level. Keeping the same average power in all cases means that the PA saturated power must correspondingly increase.

Figure 3.1 shows that if the signal PAPR is 6 dB, then the PA must be designed to provide a power at least four-times higher than the signal average power needed in the communications system specification. This factor assumes that the PEP is equal to the amplifier's saturated output power P_{SAT}, meaning that the PEP has no OBO from P_{SAT}. For linearity reasons, the PEP must remain below P_{SAT}; there is some OBO and the PA must be designed for a correspondingly higher value of P_{SAT}. Similarly, if the PAPR is 13 dB, as it can be for OFDM signals [3], then the PA must be designed to provide 20-times the power needed in the communications system specification. It quickly becomes apparent that the signal PAPR value has a large impact on the implementation cost of a wireless communication system.

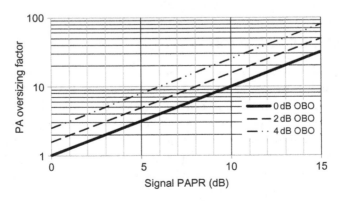

Figure 3.1 PA design scaling factor based on the signal PAPR value.

It is interesting to note the progression of signal PAPR values over time as system specifications evolve. This is shown in Figure 3.2. Older signals used in GSM, Bluetooth 1.0, and ZigBee are constant envelope (CE), so the PEP and the average power are identical. This is ideal for minimum cost and maximum PA energy efficiency. However, this is not ideal for bandwidth efficiency (bits per second per hertz of occupied bandwidth). As the bandwidth efficiency increases, physics necessitates that the PAPR must increase, and this is seen in the EDGE, UMTS, and HSPA signals. However, this pattern is broken with the OFDM signal family, which includes LTE, where the PAPR continues to increase while the bandwidth efficiency does not [3].

In addition to PAPR, the signal order is important because it guides the PA designer in the amount of waveform distortion that the transmitted signal will tolerate. The signal order (M) shows us engineers how many different waveforms the signal has that each carry different information. The higher the value of M, the more different waveforms

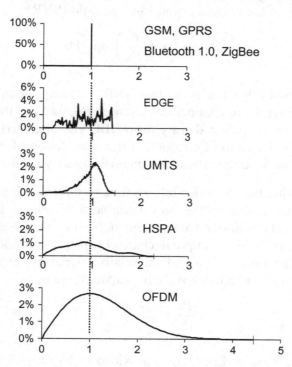

Figure 3.2 Envelope probability densities showing the progression of signal PAPR values in (mostly) the cellular communications standards, normalized to signal average power of 1.0. Here, the envelope voltage values are shown. The PAPR values are the square of these envelope values; for OFDM, the PAPR is $4.4^2 = 19.4$ (13 dB).

that the signal modulation has that carry unique information. For a binary digital modulation, the general formula for signal order is:

$$M = 2^b \tag{3.1}$$

where b is the number of bits carried in each signal symbol. For QPSK, $b = 2$ so $M = 4$. For 64-QAM, $b = 6$ and $M = 64$. And for OFDM with 52 subcarriers, each subcarrier modulated with QPSK, $b = 52 \times 2 = 104$ and $M \sim 10^{31}$! Not surprisingly, QPSK is quite tolerant of signal distortion, 64-QAM is less so, and OFDM requires very high linearity in its transmitter (and receiver). With such a high signal order, this signal is not tolerant of waveform distortion.

Increasing operating bandwidths is another difficulty for PA design. Much of this follows from the channel information capacity work of Claude Shannon, who in 1948 published that given perfect signal processing and coding, the capacity density of any noise-limited channel measured in bits per second per hertz of bandwidth is bounded strictly by a function of the present signal-to-noise ratio (SNR):

$$\frac{C}{B} \leq \log_2 \left(1 + \frac{P_S}{P_N} \right) \text{ bps/Hz} \tag{3.2}$$

Note carefully that this is the true SNR, meaning the ratio of channel signal power P_S to channel noise power P_N, and *not* the individual bit energy-to-noise power density ratio (IBEND, often written E_b/N_0), which is instead a ratio of energies. It therefore follows from Eq. (3.2) that if the SNR is limited, then the channel capacity is also limited.

Additionally, Eq. (3.2) also tells us that no matter how sophisticated our signal processing is, the only variable available to improve the channel capacity is SNR. Thus, in our noise-limited systems, the only real variable available to improve channel capacity is to increase signal power. Solving Eq. (3.2) to find the relative signal power necessary to achieve a particular channel available capacity, we get:

$$\frac{P_S}{P_N} = 2^{(C/B)} - 1 \tag{3.3}$$

A graphic view of Eq. (3.3) is provided in Figure 3.3. This clearly shows the exponential nature of Shannon's capacity limit when viewed from the corresponding SNR requirements.

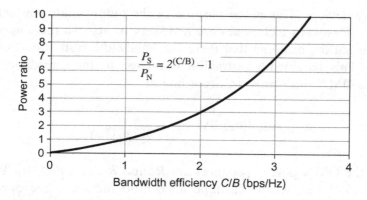

Figure 3.3 Relative power required by Eq. (3.3) for desired bandwidth efficiencies. To get 1 bps/Hz, one unit of power is needed. Seven units of power are needed to increase the available bandwidth efficiency to 3 bps/Hz. To get to 10 bps/Hz, 1023 units of power are needed.

For practical channels the effective bandwidth efficiency limit is 4 bps/Hz, and it can be increased to near 6 bps/Hz with significant increases in cost. To get high on-air data rates, the only remaining option is to use more bandwidth for the signal. For example, to send 100 Mbps with a signal providing 4 bps/Hz, the signal requires 100/4 = 25 MHz bandwidth. Similarly, for a goal of 1 Gbps, this signal needs to expand to 250 MHz bandwidth. This is difficult to achieve, particularly in mobile devices. However, this is exactly the path taken by the standards bodies 3GPP and IEEE 802.11.

Additionally, mobile devices are severely power-limited. Battery voltage available is also small, nominally 3.6 V for modern lithium-ion cells. Ohm's law says that to generate power from a fixed voltage, the load resistance must be small. For a modern mobile phone, this load resistance is approximately 3 ohms, a very low value. If the signal PAPR is high, then the design load resistance must be reduced further to enable the PA to generate the necessary PEP from the same limited power supply. In all cases, the RF transistor must be capable of providing an output resistance far smaller than this load resistance to assure that nearly all of the output signal provides power to the load and is not dissipated in the transistor itself. This means that the RF transistor must be increasingly larger as the load resistance decreases, whether driven by increased PAPR or simply more power for increased communication range.

Whatever the design load value for the RF transistor turns out to be, the impedance must be converted to 50 ohms before connecting to other devices. This is the purpose of the output matching network (OMN). Resonant techniques are necessary to achieve this impedance transformation, meaning that there is a bandwidth restriction from the OMN. This bandwidth restriction is related to the impedance ratio that the OMN must provide, according to:

$$\text{Amplifier BW} = \frac{f_{\text{signal}}}{\sqrt{(R_2/R_1) - 1}} \tag{3.4}$$

where the OMN port impedances are R_1 and R_2, and $R_2 > R_1$. We see in Eq. (3.4) that when the impedance ratio increases, the operating bandwidth decreases. This is a major difficulty for PA design when wideband signals having large PAPR values are used.

Standards organizations today have difficulty finding enough bandwidth in spectrum released by government regulators to meet the total signal bandwidth desired to meet the data rate goals of wireless signals. Their answer to this problem is predictable and causes yet another major problem for PA design. The approach is called carrier aggregation, which essentially means that if one bandwidth allocation is insufficient to form a signal that meets the data rate goals, then divide the data across several different and separate signals that are placed in separate frequency allocations, coordinating the data across all of these separate signals such that the total data rate goal is met. This is very easy to say, but nearly impossible to actually do.

The problem is amplifier bandwidth, compounded by amplifier linearity demands. When multiple (N) signals are used, assuming that each experiences identical path loss, they *each* must have the same average power to have equivalent communication range performance. This increases the PA average output power by a factor of N. From the PA's point of view, the N signals are actually just pieces of one very wideband signal. Ignoring for the moment that having multiple signals may cause an increase in overall PAPR, and therefore also PEP, for a PA operating from a battery (or other fixed voltage), the presence of carrier aggregation forces the output load resistance to be further reduced by this factor of N.

According to Eq. (3.4), a smaller value of transistor load impedance R1 reduces the available bandwidth from the impedance transformation (matching) network. This is exactly opposite to the wider

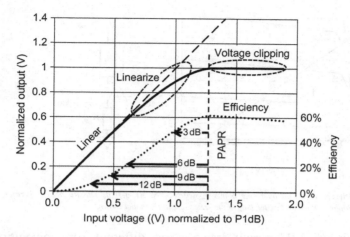

Figure 3.4 Energy efficiency of an amplifier is maximum around the onset of output clipping. For linear signals that cannot tolerate much distortion, the PEP must not exceed this clipping level. Then, as signal PAPR increases, the available energy efficiency decreases.

bandwidth the PA must have to support multiple signals in separate bands. Physics is clearly not our friend in actual implementation of any carrier aggregation system. This usually means that even if a way is found to meet these operating requirements, the cost will be high. Nothing comes for free.

Power supplies and heat sinks are a significant cost for any transmitter. Maximizing cost effectiveness therefore means that the transmitter must operate at the maximum possible energy efficiency: for the required output RF power, the power supply must be small and there must be only a small amount of heat given off that needs to be removed in the heatsink. This is actually a tremendous problem, as seen in Figure 3.4, particularly when the signal PAPR is large (more than 4 dB).

3.3 A REVERSE DESIGN APPROACH

Normally, we start by creating a linear circuit design and then work to improve its energy efficiency. Instead, this reverse approach starts with a very energy-efficient circuit—a switch—and then proceeds to make it generate signals accurately. To the outside, the signal quality appears as if the circuitry is operating with very good linearity, although actually the circuitry has all linearity suppressed.

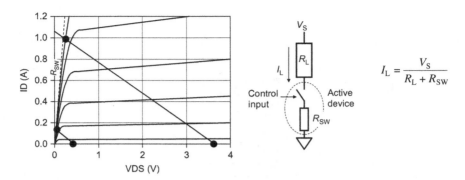

Figure 3.5 Division of circuit operating points in switch-based RF design. The RF device operates only at the extreme positions of its load line, leading to the series resistance model shown.

Switch-based RF circuit design is very different from conventional linear circuitry [2]. One key aspect is seen in Figure 3.5. Here, we see that the circuit has two operating points along the load line: one when the device is OFF and the current through it is zero, and the other when the device is ON and the current is the maximum available and the voltage drop is small but not zero. This leads to the circuit model of Figure 3.5, where the transistor is shown as a series combination of an ideal switch with a fixed resistor R_{SW} representing the transistor ON resistance. The load R_L is in series with the transistor switch, and the current I_L flowing through the load is readily written according to Ohm's law as shown in Figure 3.5.

Small signal gain has no meaning in this circuit, because the drive signal is always large enough to get the device resistance very close to R_{SW}. Under this condition, the device characteristic curves are close together, meaning that the ON resistance changes very little with changes in the drive signal. This is another way of saying that device transconductance is very small when it is ON, a result that will be very important in the later discussion of circuit stability.

Evaluating the circuit model of Figure 3.5, we get a surprise. When the RF device is a bipolar transistor, the ON resistance projection misses the $I-V$ origin when device current is zero. This offset voltage is related to an offset voltage V_{AMO} (amplitude modulation offset voltage) that we need to add to the model from Figure 3.5, as shown in Figure 3.6. It is suspected that the value of V_{AMO} is related to the $V_{CE,sat}$ of the bipolar transistor, but this has yet to be proven. The actual process used to determine the value of V_{AMO} is presented later after the discussion of amplifier "gain" when operation is nonlinear.

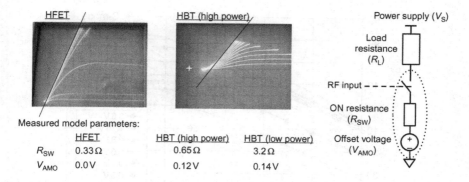

Measured model parameters:

	HFET	HBT (high power)	HBT (low power)
R_{SW}	0.33 Ω	0.65 Ω	3.2 Ω
V_{AMO}	0.0 V	0.12 V	0.14 V

Figure 3.6 Evaluating the parameters for this switch-based circuit model for typical HFET and HBT transistor types. The FET ON resistance projection passes through the origin, but the HBT projection of its ON resistance shows an offset away from the origin.

3.3.1 PA Operating Modes

All PAs actually have two inputs, the RF signal input and the power supply input. Usually, we operate PAs at a constant power supply value, allowing us to ignore this as a separate input. However, some very interesting results appear when a PA is evaluated as a three-port (two inputs, one output). One representative sample of this two-parameter sweep is shown in Figure 3.7.

Figure 3.7A shows an overlay of seven curves, with each one measuring the PA output power across a sweep of input power for a particular value of power supply voltage. In the section labeled "L-mode," this amplifier outputs the same power even as the power supply varies by nearly 3:1. This is exactly what conventional linear amplifier theory teaches us, that to first order the output signal from a linear amplifier does not depend on the value of the power supply. To the right side of Figure 3.7A, we see the region labeled "C-mode," which stands for compressed operation. Here, the output power does not change with varying input power, but rather changes significantly with varying power supply.

Finally, at the lower left of Figure 3.7A is a region labeled "P-mode," which stands for product operation. Here, the output power changes with both the input power and the power supply. This operating mode only exists at very low values of the applied power supply, generally <0.5 V. Because PAs are usually never operated at such low power supplies, this mode is relatively unknown. This is not another linear operating region, because the transistor is not operating as a

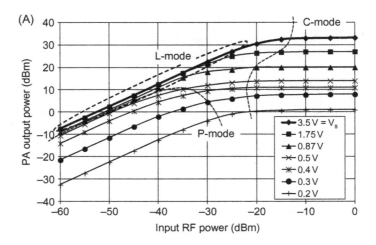

Figure 3.7 Operating mode of a PA varies with the value of input RF power and the applied power supply voltage: (A) an overlay of curves with power supply varying across 0.2–3.5 V showing the three possible operating modes and (B) general characteristic comparison among these three modes.

controlled current source (CCS), as is required of conventional amplifier biasing. Rather, the transistor is operating as a variable resistor with two independent control ports [4]. In this region there can be significant distortion on the output signal, so it is very rarely used. We also observe from the chart that the gain is low (much lower output power for the same input power) while the sensitivity of output power change per decibel of input power change remains 1:1.

The chart in Figure 3.7B provides a summary of these three operating modes, sorted according to which input parameter variation the output power value is sensitive to. If the output power is sensitive to varying input power and not sensitive to a varying power supply, then the amplifier is in conventional linear CCS operation. If the sensitivities are swapped, so that the output power is not sensitive to varying input power and is instead sensitive to a varying power supply, then the amplifier is in compressed operation (C-mode). When output power is sensitive to both varying input power and to a varying power supply, then the amplifier is in product operation (P-mode).

3.3.2 What Does "Gain" Mean When Nonlinear?

The concept of amplifier gain seems intuitively obvious: the output signal changes in proportion to changes in the input signal. When one reviews the past century of amplifier literature, there are actually two definitions used to measure gain: the first derivative of the amplifier voltage transfer function and the ratio of the output signal divided by the input signal. The first gain definition is called the *slope* gain, and the second definition gives the *ratiometric* gain. Fortunately, when the amplifier is operating linearly, both of these measures provide the same answer [5]. And for this reason the electronics industry has gotten away with being lazy and not discerning carefully between these two gain measurement methods.

When the amplifier is operating nonlinearly, these two gain measures diverge and can differ greatly. For example, in C-mode there is no output change with an input change, so the slope gain is zero. But there is still more output power than input power, so the ratiometric gain is not the same. This is shown graphically in Figure 3.8. It is important to remember that both measures of gain are correct in that they both examine aspects of the same physical amplifier. To remove ambiguity it is essential to explicitly specify which gain measure is being used. By far the usual measure of amplifier gain is the ratiometric type.

Why this is a problem is shown in Figure 3.9. For an ideal (hard limiting) amplifier, the ratiometric gain begins to decrease at the onset of clipping, when the output waveform becomes distorted. However, the output power continues to increase, even though the output waveform is no longer increasing in amplitude. Of course, this is due to the fact that the fundamental frequency power of a squarewave is approximately 2 dB larger than the power of a sinewave of the same amplitude. So, while the last 2 dB of output power is being generated, the output waveform is being increasingly distorted from a sinewave to a squarewave. Thus, we revisit Figure 3.7A and understand that in C-mode, this is not the onset of clipping but rather the completed transition of the

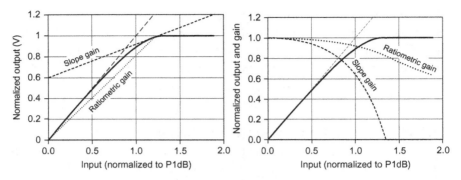

Figure 3.8 *The two measures of gain diverge greatly when an amplifier operates nonlinearly. When the output stage clips, the slope gain goes to zero. The ratiometric gain never goes to zero as long as there are finite input and output powers.*

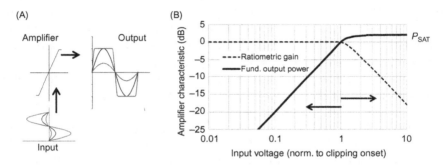

Figure 3.9 *For an ideal linear amplifier, linearity (slope gain) is maintained all the way to clipping: (A) waveform distortion mechanism from increasing clipping and (B) output power increases and g_r decreases with harder clipping.*

input sinewave into a squarewave. The amplifier output waveform at saturated output power is not sinusoidal but is nearly square.

Now we can complete our definition of V_{AMO}. The actual definition of V_{AMO} comes from C-mode measurements of amplifier output power versus variable power supply. Knowing that the output waveform is square in this mode, the output power must follow a square-law relationship as the power supply is varied, with specific values dependent on the PA load resistance (R_{PA}) and transistor efficiency parameters collected within the scaling factor α. Figure 3.10A shows an amplifier that matches this square-law model quite accurately. Figure 3.10B shows an amplifier that requires an offset to be applied to the power supply before evaluating the square-law relationship for a good model fit. This required model offset is the definition of V_{AMO}.

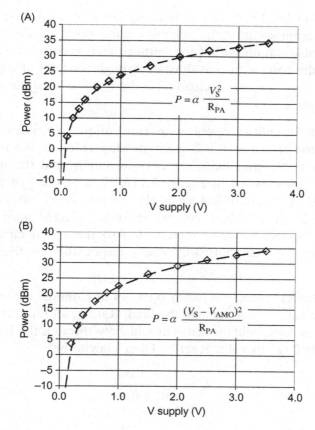

Figure 3.10 Defining the model parameter V_{AMO}: (A) output power versus power supply for an amplifier with $V_{AMO} = 0$ and (B) output power versus power supply for an amplifier with nonzero value (120 mV) for V_{AMO}, defined as the voltage offset necessary into the square-law model providing a best fit to the actual PA data.

In general, it is observed that V_{AMO} is essentially zero for amplifiers using FET devices. When an amplifier is implemented using bipolar devices, whether BJT or HBT, a nonzero value for V_{AMO} is always observed. How this is related to the offset seen in the measurements of Figure 3.6 and/or to the known $V_{ce,sat}$ characteristic of bipolar transistors is not yet theoretically established. Such a result will be welcome in the C-mode amplifier design community.

3.3.3 Apparent Linearity: Output Signal Accuracy

From Figure 3.4 we observe that the energy efficiency of an amplifier is maximum when the output stage is clipping, which means that we desire to operate in C-mode when an operating goal is high energy efficiency. But there is no circuit linearity in C-mode operation, so output signal accuracy must be achieved using other means. Here, the important observation is that the output signal accuracy is dependent on having the correct current flowing through the load, irrespective of what the transistor may actually be doing. The analysis of this is shown in Figure 3.11.

When the amplifier operates in conventional CCS linear mode, the transistor *regulates* the current flowing through the load. This current has two components, the bias current I_Q and the signal current I_{SIGNAL}, as shown in Figure 3.11A. The product of the signal current and the load resistance is the output signal, so we get the design condition that bias current must be selected such that the transistor provides a linear relationship to the signal current from the input signal. The transistor is fully responsible for the amplifier linearity.

When the RF power transistor operates in C-mode, it cannot regulate the current flowing through the load. Rather, all that the RF transistor does is select whether current will flow through the load. When current does flow, its value is set by Ohm's law to be:

$$I_L = \frac{V_S - V_{AMO}}{R_L + R_{SW}} \tag{3.5}$$

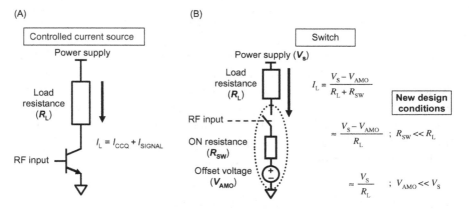

Figure 3.11 Amplifier models used to evaluate output signal linearity: (A) operated as CCS, the transistor regulates the signal current through the load and is therefore fully responsible for output signal accuracy; (B) in C-mode, the current through the load is set by the value of the power supply, transferring responsibility for output signal accuracy from the RF transistor to the external power supply.

Here, we encounter two design conditions for C-mode power stages. First, select the RF transistor such that its ON resistance is negligible when compared with the load resistor value. This gets the load current to be approximately:

$$I_L \cong \frac{V_S - V_{AMO}}{R_L} \tag{3.6}$$

Reviewing Eq. (3.6), it is also desirable to have V_{AMO} to also be negligible compared with the power supply. This situation is realistic when either a transistor is selected that exhibits $V_{AMO} = 0$ or the power supply is maintained at a high enough value to assure that any nonzero value of V_{AMO} remains negligible. Then, the load current is effectively described by:

$$I_L \cong \frac{V_S}{R_L} \tag{3.7}$$

The situation described by Eq. (3.7) is exceedingly important. We observe that the only parameters are the value of the power supply and the transistor load resistance. Both of these are external to the RF transistor, showing that a properly designed C-mode stage does not have any RF transistor characteristics important (to the first order) for setting the signal current. Unlike any linear amplifier, this transistor has no responsibility for output signal envelope accuracy. This responsibility for output signal envelope accuracy is transferred primarily to the dynamic power supply (DPS).

3.3.4 Stage Series Resistance

It is necessary to assure that the RF transistor is actually operating in C-mode. This is preferably performed using external measurements with simple instruments using the concept of stage series resistance (SSR). The definition of SSR is:

$$SSR \equiv \frac{V_S}{I_L} \tag{3.8}$$

which says that the only necessary measurements are the stage voltage V_S and its current draw I_L. According to Eq. (3.7), this SSR is actually a measurement of the load resistance when the stage is operating in C-mode. Thus, C-mode operation is confirmed when the SSR measurements plot as a horizontal line on a graph like that shown in Figure 3.12.

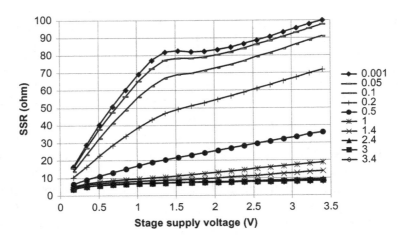

Figure 3.12 Operating modes, whether CCS (linear) or C-mode (switching), are readily identified on a plot like this of SSR versus the applied power supply voltage. Flat horizontal lines identify C-mode operation; lines with a slope identify CCS operation.

The amplifier providing the data in Figure 3.12 does not always plot as a horizontal line. To understand what is happening here, it is useful to rewrite Ohm's law as:

$$R = \left(\frac{1}{I}\right) V \tag{3.9}$$

which has the form of a simple line with slope of the reciprocal of the current. In Figure 3.12, there are several curves that have straight slopes, which show that in these measurements the transistor is operating as a CCS, sometimes with changing values of bias current.

3.4 OUTPUT POWER CONTROL

Transmitters have more tasks than making the desired signal with sufficient quality. In most wide area communications systems, they also must keep this signal quality (and all the corresponding envelope dynamic range) while controlling the output power across a separate dynamic range. For GSM, this power control dynamic range is approximately 40 dB, whereas for CDMA systems such as WCDMA this power control dynamic range exceeds 70 dB, having output power requirements down to −50 dBm. Whatever the technology used, these requirements must be met.

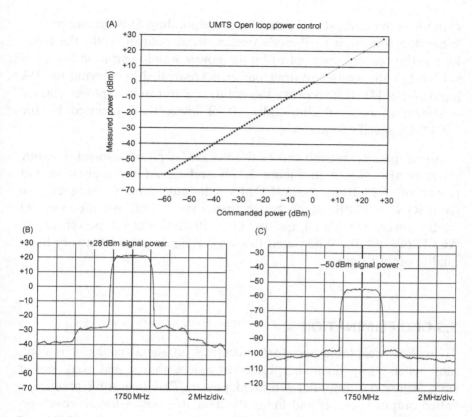

Figure 3.13 Output power control across +28 to −60 dBm of a nonlinear transmitter generating a WCDMA signal: (A) measured output power (dots) compared with the ideal requirement (line), (B) output signal quality at +28 dBm WCDMA output power, and (C) output signal quality at −50 dBm WCDMA output power.

Examining Figure 3.7A for low output power, we note that C-mode appears to have a problem getting to the low output power of WCDMA. In contrast, a linear amplifier has no difficulty getting to low output power—one simply reduces the input power. We also note that P-mode can provide the needed low output power. One solution is to mode-switch between C-mode at high output powers and P-mode at low output powers. This technique is used in the measured WCDMA transmitter data in Figure 3.13.

For the top 30 dB of output power control, C-mode is used. This means that all envelope variations and control of average output power are provided by varying the power supply in accordance with Eq. (3.7). PA input power is held sufficiently high to assure that

C-mode is maintained, validated by corresponding SSR measurements. When the transition to P-mode occurs, these controls split: the envelope variations remain applied by the power supply variation (at levels <1 V), but the average output power is controlled by varying the PA input power [4]. Just to prove the point, this transmitter power control is shown to be −60 dBm, fully 10 dB lower than required by the WCDMA specification.

Signal quality at each end of the WCDMA power control dynamic range is also shown in Figure 3.13B and C. At the highest output power of +28 dBm, the WCDMA adjacent channel leakage ratio (ACLR) is −49 dB, much better than the −33 dB specification. At 78 dB lower, the signal quality at −50 dBm output power shows ACLR of −44 dB. When used this way, P-mode provides a very high-quality output signal.

3.5 OBO ELIMINATION

To realize acceptable waveform accuracy on signal peaks, it is common when using linear amplifiers to assure that signal peak power stays below the amplifier-saturated power. This operating practice is called output back-off and limits the available power for an envelope-varying (EV) signal from the amplifier to:

$$P_{EV} \leq P_{SAT} - \text{PAPR} - \text{OBO} \tag{3.10}$$

For the C-mode transmitter, the RF transistor is always operating at its available saturated output power. This negates the need for any OBO; any OBO is not possible in a C-mode transmitter [6]. Thus, the maximum available PEP is P_{SAT}, and the available power becomes:

$$P_{EV} \leq P_{SAT} - \text{PAPR} \tag{3.11}$$

which is higher than from Eq. (3.10). We conclude that for the same size of RF power transistor, more power is available from a C-mode transmitter than a linear transmitter because of the elimination of all OBO limits. This results in the situation shown in Figure 3.14, where the envelope peaks of the EV signal bursts are at the same level as the saturated output power used for the CE signal.

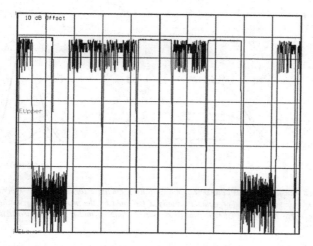

Figure 3.14 A C-mode transmitter is always operating at the available saturated power, so no OBO conventionally used with linear amplifiers does not apply. PEP of an EV signal matches the saturated power used for CE signals.

3.6 STABILITIES: CIRCUIT, THERMAL, AND MANUFACTURING

For any manufacturable product, it is essential that several stabilities are assured. For electronic amplifiers, there are three important stabilities: stability of the circuit, stability over temperature, and stability across manufacturing [7]. All three of these stabilities are required in order to achieve low cost and uniformity of performance. These are critical considerations that must happen, above and beyond the design, to achieve basic application performance.

3.6.1 Stability of the Circuit

Circuit stability is simply the characteristic of not oscillating. We require the output of any circuit to be dependent only on the circuit inputs and nothing else. For an amplifier, this means that the Barkhausen criteria (for oscillation), which comprise a voltage gain ($A_V = g_m R_L$, the slope gain) of unity along with a net phase shift ($\Delta\phi$) of an integer multiple of 2π:

$$A_V \geq 1$$
$$\Delta\phi = 2\pi k, \quad k \text{ an integer}$$

must never be satisfied. Because linear amplifier gain cannot be compromised, this means that signal phasing is the only available degree of design freedom for circuit stability in a linear amplifier. This is often worked out

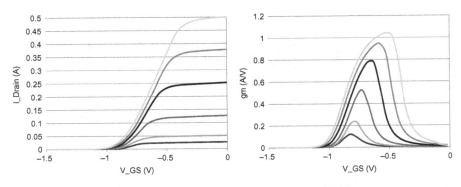

Figure 3.15 Device transfer functions and corresponding transconductance for a transistor. Operating in C-mode, the transconductance is very low. Slope gain, nominally determined by the product $g_m R_L$, is therefore also very small, well below unity. This is not enough gain to support oscillation, and a C-mode circuit is unconditionally stable.

with s-parameters and stability circle analysis. Assuring circuit stability in design is more difficult as operating bandwidth is increased.

The fully compressed C-mode operation essentially removes the gain, which is important to the Barkhausen criteria. Slope gain for C-mode is effectively zero in this operating condition, as shown in Figure 3.15 [2]. This means that design considerations for stability of a linear amplifier are eliminated as part of C-mode transmitter development. All of the corresponding time and engineering resources are saved.

3.6.2 Stability Over Temperature

Everything changes its characteristics as temperature changes. This is particularly evident with linear amplifiers and the fact that temperature is a significant parameter in the transistor current regulation equations. Because the transistors used are biased to provide a particular quiescent current, and to regulate that current, any parameter that impacts how this bias impacts the quiescent current is a source of temperature dependence.

The idea of quiescent current has no meaning in the fully compressed output stage of a C-mode transmitter. Clearly, the RF transistor here does not regulate the current that flows through it. The amount of current that flows through this transistor and the load are controlled by external design. This calls for new design methods; the primary one Eq. (3.6) calls for the transistor ON resistance to be negligibly less than the intended load resistance. Thus, when the transistor ON resistance characteristic changes with temperature, as it will, the change in load

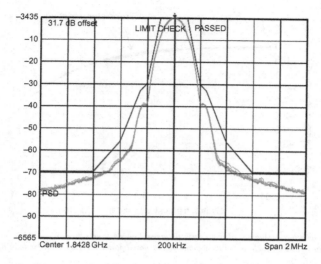

Figure 3.16 Thermal stability inherent in a C-mode transmitter, here demonstrated as an overlay of nine EDGE signal power spectral densities measured across the temperature sequence +30°C, +40°C, +50°C, +60°C, +70°C, +60°C, +50°C, +40°C, +30°C. This transmitter is developing 20 W at 1980 MHz [4].

current—and, hence, the output signal—is negligibly small. Therefore, simply by operating in C-mode, temperature stability is assured without need for testing, calibration, or temperature compensation.

Success of this approach is presented in Figure 3.16, which shows an overlay of nine measurements of EDGE signal spectra from a 20-W C-mode transmitter at 1980 MHz. This transmitter is measured through the temperature sequence of 30°C, 40°C, 50°C, 60°C, 70°C, 60°C, 50°C, 40°C, 30°C. Spectral overlap is almost perfect, except for a small region 65 dB down at +200 kHz from the signal center frequency. Still, this variation is well below the spectral mask limit, which makes it a nonissue for production.

3.6.3 Stability Across Manufacturing

It is well-known that each unit made through a manufacturing process has slightly different characteristics. These characteristic differences usually require calibration during manufacturing to assure consistent performance of product. This is the concept of manufacturing stability and is independent of the temperature stability issues discussed.

For the C-mode transmitter manufacturing differences of ON resistance, the dominant parameter of interest, appear in the design process just as resistance variation due to temperature. Testing shows that once

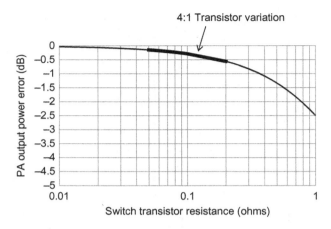

Figure 3.17 Manufacturing stability of a properly designed C-mode transmitter. When the transistor ON resistance varies by an octave on each side of its nominal value, output power variation is within 0.5 dB.

thermal stability is achieved, as long as the part variations of ON resistance are smaller than the temperature variation, no problems across part variation in manufacturing are seen [6]. Output power performance is very consistent, with one example presented in Figure 3.17.

In Figure 3.17, the part-to-part variation is allowed to vary by an octave on either side of the nominal 0.1 ohm ON resistance, for a range of 0.05−0.2 ohms. Here, the load resistance is $R_L = 3$ ohms. We note that across this 2-octave variation range in ON resistance, the C-mode transmitter output power stays within 0.25 dB from its nominal value. Again, without any compensation or control circuitry, this is inherent performance for this architecture.

3.7 AGING

As a transmitter operates over time, its characteristics may change from their original values. Such a drift in performance is called aging, and it is generally undesirable. Much of this slow performance drift is driven by power dissipation in the RF transistors. Here, the by-product of C-mode operation is a reduction in dissipated power. This must correspond in lower operating temperatures and therefore also improved stability over time.

Measurements from another C-mode transmitter operated continuously at +85°C for 1000 hours are shown in Figure 3.18. Performance

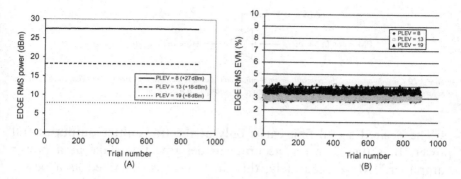

Figure 3.18 Measured data from high-temperature aging tests: (A) output power consistency at high, middle, and low specified powers and (B) error-vector magnitude (EVM) consistency at the same output power levels (PLEV is the command name for setting the transmit power level).

measurements were recorded at intervals slightly exceeding 1 hour. Consistency of RF output power for EDGE (3π/16-8PSK) is presented in Figure 3.18A, where there is no discernible change across the 42-day test. An overlay of the modulator distortion at these same output powers is shown in Figure 3.18B. Again, extremely consistent performance is measured all through the extended test. Stability of this C-mode transmitter over extended time is excellent.

3.8 CATEGORIZING C-MODE OPERATION

C-mode operation has properties that are different from conventional linear amplifiers. Is there more we can say about this operating mode? To answer this question, we revisit signal processing basics. A linear signal process has two ports, an input and an output. The transfer function between these ports, such as V_{out}/V_{in}, has the largest power series expansion coefficient as the linear (first-order) one. For comparison, in a frequency doubler or squaring circuit, the second-order coefficient dominates.

By definition, a polar signal process has three ports, two inputs and one output [8]. One input signal controls a polar parameter (magnitude or phase) of the signal passing from the second input to the circuit output. Returning to Figure 3.7, we note that varying the power supply of an amplifier operating in C-mode directly controls the output signal magnitude. Because magnitude is one of the polar coordinates, a C-mode stage is *by definition* a polar signal processor. This is explicitly shown in Figure 3.19.

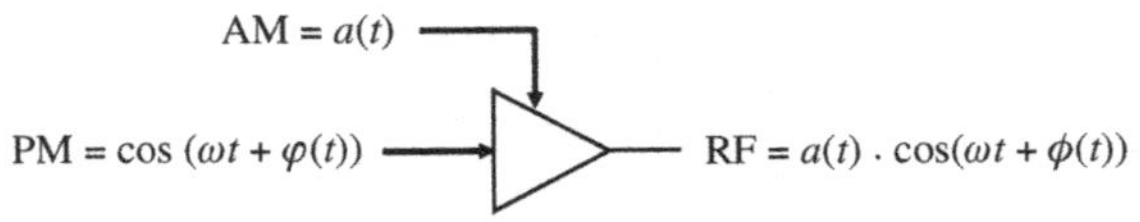

$$AM = a(t)$$

$$PM = \cos\left(\omega t + \varphi(t)\right) \longrightarrow \triangleright \longrightarrow RF = a(t) \cdot \cos(\omega t + \phi(t))$$

Figure 3.19 C-mode operates the transmitter final amplifier as a three-port, not as a two-port. This corresponds to a polar modulation process, with the final power stage operating as a multiplier.

The desired signal first exists only at the transmitter output at full power. It is therefore not accurate to describe this stage as a power "amplifier." More accurately, this final stage is described as a power modulator. It is a key component to the output stage of the envelope elimination and restoration (EER) technique [9–11].

This multiplication operation is directly analogous to signal mixing. One could then name this transmitter structure as "Mixer Last" with justification. Another interpretation of the process shown in Figure 3.19 is of the RF signal sampling the envelope signal, with the desired output taken at the frequency of the first alias. As long as we generate "narrowband" signals, meaning that the RF frequency is much greater than the bandwidth of the envelope signal, this interpretation is useful.

It is important to emphasize that C-mode is not an envelope-tracking (ET) operation. ET is not a polar operation in that the amplifier remains linear in ET [11,12]. As such, ET operation necessarily provides lower energy efficiency than C-mode operation [8]. When ET designs are optimized for improved energy efficiency, all known instances actually operate as C-mode polar modulators, at least at larger signal magnitudes.

3.9 CONCLUSION

Changing the design approach from the conventional "design a linear circuit and then find ways to make it more efficient" to the reverse, "start with an efficient circuit and make it behave as if linear" does work. This maximally efficient circuit is a switch, resulting from operating the RF transistor in compressed mode (C-mode). This switch action has no circuit linearity at all. It is no surprise that the conventional linear circuit design practices do not apply to this reverse approach. In particular, the use of s-parameters does not apply because of the circuit linearity assumptions inherent in the definition of s-parameters.

Without circuit linearity, what is meant by the term "gain" needs to be carefully examined. It is found that the two definitions for gain that are traditionally used interchangeably, namely slope gain and ratiometric gain, no longer provide the same results when the transistor operates in C-mode. Slope gain goes to zero, but ratiometric gain does not. This explains how there can be greater output power than input power, an important gain concept, even when circuit linearity is completely suppressed.

Design procedures that do apply to C-mode design are developed, which focus around three steps:

1. Characterize the RF device for C-mode operation, particularly quantifying the model parameters R_{SW} and V_{AMO}.
2. Select the RF device to assure that R_{SW} is negligibly small compared with the intended load resistance the transistor must work into.
3. Operate the power supply high enough that the value of V_{AMO} is negligible by comparison.

Successfully implementing these three steps provides a C-mode transmitter that exhibits several inherent stabilities that are important for fast design, consistent manufacture, and very consistent long-term performance.

Measurements of C-mode transmitters demonstrate that there are four innate stabilities in this architecture. Circuit stability is assured because there is no slope gain in this circuit. This guarantees that the Barkhausen criteria cannot be met: a switch cannot oscillate. This stability is assured only as long as the circuit remains operating in C-mode. Fortunately, this can be easily shown using SSR measurements.

Stability of output power across temperature results from successfully implementing design step 2 mentioned previously. All transistors change their characteristics with varying temperature, but as long as this transistor variation is very small when compared with the external load resistance, the actual signal current will be effectively unchanged.

Stability of output power across transistor instance (manufacturing) variations is realized through the same procedure. Whether the transistor ON resistance varies through temperature changes or by device to device, the step from Eq. (3.5) to Eq. (3.6) still holds true.

The remaining variation is the accuracy of each instance of the load resistance, which traditionally has smaller variance than any semiconductor process.

Finally, stability of long-term performance follows from the initial premise of this design: begin with the circuit having greatest energy efficiency. Good energy efficiency necessarily means that the dissipated power is lower for the same output power. Lower dissipated power means that increase in temperature is reduced, improving both reliability and long-term performance consistency. Additionally, these innate stabilities remove design steps for circuit stability and temperature compensation, and can reduce manufacturing line calibrations. Streamlining development and manufacturing processes lowers cost, an ever-present objective in communications engineering.

C-mode transmitters still present challenges beyond the very different design procedures. The main challenge is also evident in Figure 3.7, where the output power is still large even when the power supply is 0.2 V. It is difficult to get the output power from a C-mode transmitter to go to zero. Adoption of signals that have zero-crossings in their vector diagrams poses difficulty for C-mode transmitter design. However, adopting signals that do have finite envelope floors (e.g., EDGE, π/4-QPSK) are readily amenable to taking advantage of C-mode transmitter properties.

REFERENCES

[1] Cripps SC. RF power amplifiers for wireless communication. 2nd ed. Boston/London: Artech House; 2006.

[2] McCune E. Polar modulation and power amplifiers. Workshop WSC at the 2009 international microwave symposium, Boston; June 7–11, 2009.

[3] McCune E. This emperor has no clothes. IEEE Microwave Mag 2013;48–62.

[4] McCune E. Multimode transmitters: easier with strong nonlinearity. In: van Roermund A, Casier H, Steyaert M, editors. Analog circuit design: smart data converters, filters on chip, multimode transmitters. Springer; 2010 [chapter 13].

[5] McCune E. Gain: changed meanings for compressed amplifiers. In: Proceedings of the 2013 Midwest symposium on circuits and systems (MWSCAS), Columbus, OH; August 2013.

[6] Sander W, Schell S, Sander B. Polar modulator for multi-mode cell phones. In: Proceedings of the 2003 custom integrated circuits conference (CICC), San Jose; September 2003.

[7] McCune E. High-efficiency multi-mode, multi-band terminal power amplifiers. IEEE Microwave Mag 2005;6(1):44–55.

[8] McCune E. Envelope tracking or polar—which is it? [Microwave Bytes]. IEEE Microwave Mag 2012.

[9] Kahn LR. Single sideband transmission by envelope elimination and restoration. Proc IRE 1952;40(7):803—6.

[10] Raab FH. Intermodulation distortion in Kahn-technique transmitters. IEEE Trans Microwave Theory Tech 1996;44:2273—8.

[11] Raab FH, Asbeck P, Cripps S, Kenington PB, Popovic ZB, Pothecary N, et al. Power amplifiers and transmitters for RF and microwave. IEEE Trans Microwave Theory Tech 2002;50(3):814—26.

[12] Buoli C, Abbiatti A, Riccardi D. Microwave power amplifier with 'Envelope Controlled' drain power supply. In: Proceedings of the 25th European microwave conference; September 1995. p. 31—5.

CHAPTER 4

mmW Doherty

Mohammadhassan Akbarpour, Fadhel M. Ghannouchi and Mohamed Helaoui

Intelligent RF Radio Laboratory (iRadio Lab), Department of Electrical and Computer Engineering, University of Calgary, Calgary, AB, Canada T2N 1N4

4.1 INTRODUCTION

Millimeter-wave (mmW) frequencies (30–300 GHz) are being used for many applications in the modern world. These applications include,but not are limited to, radio astronomy, remote sensing, automotive radars, military applications, imaging, security screening, and telecommunications.

There are various telecommunication standards that specify the carrier frequency in the mmW frequency range. IEEE 802.16 specifies a frequency range of 10–66 GHz for wireless metropolitan area networks (MANs). IEEE 802.15 and ECMA-387 standards specify a frequency range of 57–66 GHz for high data rate personal area networks. IEEE 802.11ad specifies the 60-GHz frequency for wireless local area network (LAN) applications.

In these standards, spectral efficient complex modulation schemes are recommended to reach very high data rates. Spectral efficient modulation schemes like high-order quadrature amplitude modulations (QAM) and orthogonal frequency division multiplexing (OFDM) result in high peak-to-average power ratio (PAPR) for the signal to be transmitted. Conventional power amplifiers with high PAPR input signals have low energy efficiency.

There are several techniques for improving the efficiency of power amplifiers in transmitter systems with high PAPR signal. Envelope tracking (ET), envelope elimination and restoration (EER), linear amplification with nonlinear components (LINC), and the Doherty amplifier are some of the architectures that are used to address the low amplifier efficiency in the presence of high PAPR signals. Among these

Linearization and Efficiency Enhancement Techniques for Silicon Power Amplifiers.
DOI: http://dx.doi.org/10.1016/B978-0-12-418678-1.00004-0

solutions, the Doherty amplifier is the most accepted solution in wireless communications because of its simplicity and promising performance. In this chapter, the challenges of implementing the Doherty amplifiers in silicon technologies for mmW frequency applications are discussed.

4.2 DOHERTY AMPLIFIER

The Doherty amplifier was first introduced in 1936 by W. H. Doherty [1]. The Doherty amplifier is a multibranch amplifier that has higher average efficiency than conventional class AB or class B amplifiers for modulated signals. The Doherty amplifier works based on active load-pulling. The load-pulling is dependent of the amplifier's input power and is accomplished by the specific bias points and interconnection of amplifier branches.

4.2.1 Doherty Structure

Figure 4.1 shows the load lines for a class B biased FET (BJT) transistor. Before saturation, the value of efficiency for a class B amplifier only depends on the voltage swing and is independent from the load impedance. In Figure 4.1, for the voltage swing shown, all the load lines give the same efficiency but different output powers. Theoretical maximum efficiency for all load lines is $\pi/4$ (78%), and it is obtained when the voltage swing reaches its maximum value.

The load line that gives maximum output power is designated as R_{opt}. This load line corresponds to the maximum transistor's current/voltage swing. Load impedances higher than R_{opt} lead to transistor saturation at lower output power because the current swing is less than its maximum

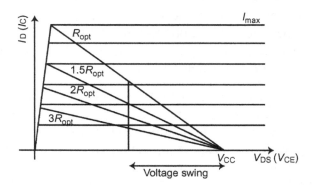

Figure 4.1 Class B amplifier load lines.

possible value. For load impedances higher than R_{opt}, the amount of maximum output power is inversely proportional to the load impedance:

$$\frac{P_{max,RL}}{P_{max,Ropt}} = \frac{R_{opt}}{R_L} \qquad (4.1)$$

It means that for higher load impedances, the transistor saturates at a lower output power and consequently reaches maximum efficiency at a lower output power. After saturation of the transistor at high load impedance, if the load impedance changes toward lower values, it can provide more output power while remaining in saturation. It means the output power is increasing while maintaining the efficiency at its peak value. The Doherty amplifier provides such load modulation to preserve the maximum efficiency over a wide range of output powers.

The structure of the Doherty amplifier is shown in Figure 4.2. The amplifier branch that is usually called the main amplifier or the carrier amplifier is an amplifier biased in class AB or class B. The second amplifier is usually called the peaking amplifier or the auxiliary amplifier and consists of a class C biased amplifier. At low input power level, the peaking amplifier is in the off-state (conduction angle is zero), and only the main amplifier branch delivers power to the load. The peaking amplifier should present high output impedance at a low input power range to avoid power dissipation into the peaking branch. At a higher input power level, the peaking amplifier has a positive

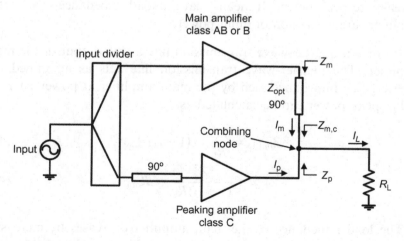

Figure 4.2 Doherty amplifier's structure.

conduction angle and delivers current to the load. By increasing the input power, the conduction angle of the peaking transistor increases and, as a result, more current will be delivered to the load at fundamental frequency. Changing the output current from the peaking amplifier leads to a change in the load impedance seen by the main branch. At the low power range, before the peaking amplifier turns on, the load impedances seen by the two branches can be calculated as:

$$Z_{m,c,\text{BO}} = \left.\frac{V_L}{I_m}\right|_{I_p=0} = R_L \tag{4.2}$$

$$Z_{p,\text{BO}} = \left.\frac{V_L}{I_p}\right|_{I_p=0} = \infty \tag{4.3}$$

At peak input power, the load impedances seen by the two branches can be calculated as:

$$Z_{m,c,P} = \frac{V_L}{I_m} = R_L\left(1 + \frac{I_p}{I_m}\right) = R_L(1 + \alpha) \tag{4.4}$$

$$Z_{p,P} = \frac{V_L}{I_p} = R_L\left(1 + \frac{I_m}{I_p}\right) = R_L(1 + 1/\alpha) \tag{4.5}$$

In Eqs. (4.4) and (4.5), α is the ratio of the peaking amplifier's output power to the main output power at peak input power. For equal peak output powers $(\alpha = 1)$, the main branch's load impedance changes from R_L to $2R_L$ as the input power to the Doherty amplifier increases to peak power. It means that the load impedance seen by the amplifiers are a function of input power.

There is a quarter-wave transmission line at the output of the main amplifier. The quarter-wave transmission line acts as an impedance inverter. The impedance seen by the main amplifier at power back-off and at peak power can be calculated as:

$$Z_{m,\text{BO}} = \frac{Z_c^2}{R_L} = (1 + \alpha)R_{\text{opt}} \tag{4.6}$$

$$Z_{m,P} = \frac{Z_c^2}{(1 + \alpha)R_L} = R_{\text{opt}} \tag{4.7}$$

The load impedance of the main amplifier decreases by increasing input power. By proper selection of the peaking amplifier's bias and transistor size, the designer can set the main amplifier to remain near saturation after peaking amplifier's turn-on, preserving high efficiency at the desired output power range.

To have proper power-combining at the output of the amplifier, the output currents of the two branches should be in-phase. To ensure in-phase output currents, a quarter-wave transmission line should be inserted at the input of the peaking amplifier to compensate for the phase shift of the main branch impedance inverter. The phase relationship can also be adjusted by using a 90° hybrid coupler as the input power divider.

4.2.2 Nonidealities in Doherty Structure

In the theory of the Doherty amplifier, the load modulation takes place at the output of the main amplifier, and the model considered for the amplifier is the ideal transistor model. It should be noted that the main amplifier includes the output matching network as well. The output matching network may alter the load modulation such that in the transistor's drain level there may be a completely different load modulation effect. However, the optimum impedances of the transistors are not real impedances, as it is assumed in the load-line analysis of the Doherty amplifier. Because of the reactive and nonlinear parasitics in the transistor, optimum impedances at back-off and at peak power are not usually multiples of each other.

The effects of the output matching network and the complex optimum impedance of the transistor are usually compensated using an offset line between the main amplifier and the quarter-wave impedance transformer (Figure 4.3). The characteristic impedance of the offset line is usually selected to be R_{opt}. With this selection, the load impedance at peak power remains unchanged while the load impedance at power back-off can be tuned to get better performance regarding power back-off in terms of efficiency, power, or linearity. The length of the offset line is determined by the actual position of the optimum impedances. Figure 4.4 shows the effect of the offset line in the performance of the amplifier.

In the theory of the Doherty amplifier, it is assumed that at low power levels the output impedance of the peaking amplifier is very high compared with the load impedance. This condition prevents the RF power from being dissipated in the peaking amplifier instead of the load impedance. Usually the output impedance of the peaking amplifier is converted to high impedance using the offset line shown in

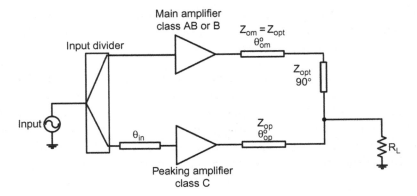

Figure 4.3 Doherty amplifier's structure including offset lines.

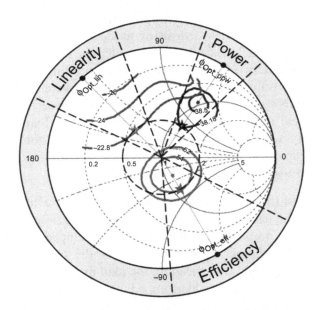

Figure 4.4 Tuning of main amplifier's back-off performance using offset line [2]. © IET 2011.

Figure 4.3 [2–4]. It can be done when the output impedance of the peaking amplifier lies very close to the edge of the Smith chart (low loss reactive impedance).

In some cases, the output impedance of the peaking amplifier is not located on the edge of the Smith chart. This is more likely to happen when the class C biased transistor's output impedance is dissipative impedance (low reflection coefficient) impedance or the circuit elements used for the peaking amplifier's output matching network are dissipative elements. At mmW frequency range, usually both of these conditions exist. As a result, the output impedance of the peaking amplifier may not be high enough to prevent power leakage into the peaking branch. In these cases, at output power back-off, a portion of the RF power is dissipated in the peaking branch and causes the efficiency to decrease at power back-off levels.

Another anomaly that can degrade the performance of the Doherty amplifier is the different behaviors of the main and peaking transistors. The bias points for the main and peaking transistors are different; as a result, in some technologies they have different characteristics in terms of AM/PM characteristics [5]. At high power levels, both amplifier branches inject current to the load impedance. To have high combining efficiency, the output currents from the amplifier branches (I_m and I_p in Figure 4.2) should be in-phase at the output combining node. By phase alignment at the input of the amplifier branches, the output current phases can be aligned only at one power level, but at other power levels the current phases may not be completely aligned. As a result, the power-combining efficiency is optimum at only one power level; at the other power levels, the output power degrades from its optimum value, resulting in a decrease in the efficiency.

4.3 mmW DOHERTY AMPLIFIERS

Although the Doherty amplifier is widely considered for mobile communications and Digital Video Broadcasting (DVB) transmitters, there are very few Doherty implementations in the mmW frequency range. Table 4.1 shows the published results obtained from Doherty amplifiers in frequency ranges of the Ku band and above. Among these published results, there are only two Doherty amplifiers (gray rows) fabricated in silicon technologies. The first mmW Complementary Metal–Oxide–Semiconductor (CMOS) Doherty amplifier implemented in 130-nm CMOS technology achieved 3% peak efficiency. The second mmW CMOS Doherty amplifier is implemented in 45-nm Silicon On Insulator (SOI) CMOS, obtaining more than 17% efficiency over the last 6-dB output power range.

Generally speaking, in power amplifier design there are two important steps for obtaining required output power efficiently. The first step is to generate the RF power efficiently, and the second step is to deliver the generated output power to the load impedance with

Table 4.1 mmW Doherty Amplifier Performance							
Reference	Year	Frequency (GHz)	Technology	Gain (dB)	P_{sat} (dBm)	PAE_{max} (%)	$PAE_{6dB\ BO}$ (%)
6	1999	17	GaAs, 0.25 μm	8	25	40	35
7	2000	18–21	InP	7.5	23	28	12
8	2000	20	GaAs, 0.15 μm	8	24	39	22
9	2007	38–46	GaAs, 0.15 μm	7	20	25	24
10	2012	22–24	GaN, 0.15 μm	15	36.8	48	29
11 (simulation)	2011	31–35	GaAs, 0.15 μm	5.5	22.8	39.2	19.8
12 (simulation)	2011	22–29	GaAs, 0.15 μm	12.5	20	32	22
13	2008	60	CMOS 0.13 μm	13.5	7.8	3	1.3
14[a]	2013	42	SOI CMOS 45 nm	8	18	21	21
15 (simulation)	2011	71–76	CMOS 90 nm	4.7	11.7	30.6	15.6

[a]Adaptive biasing is used for peaking amplifier.

minimum loss. The RF power is generated by the amplifier's active components, which are the transistors. There are limitations on the efficiency and output power performance of power transistors, depending on frequency and the transistor's technology. However, the loss-associated passive elements used at the amplifier's output for impedance-matching and power-combining degrades the amplifier's performance in terms of output power and efficiency.

4.3.1 Silicon Transistors in mmW Doherty Structure

The parameters that are used as a metric for operation of a transistor in high frequencies are the cutoff frequencies f_T (unity current gain frequency) and f_{max} (unity power gain frequency). The unilateral available power gain of a MOS transistor can be approximated as [16]:

$$G_{AU} \approx \frac{R_{ds}}{4R_g} \left(\frac{f_T}{f}\right)^2 \tag{4.8}$$

In Eq. (4.8), f is the operating frequency at which the gain is calculated. It can be seen that a transistor with higher f_T has higher gain for

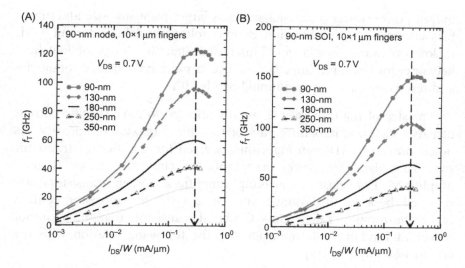

Figure 4.5 Effect of gate length and bias condition on cutoff frequency of (A) bulk and (B) SOI n-channel MOSFETs [17]. © IEEE 2006.

a given operating frequency. The factors that affect the transistor's cutoff frequency are the transistor geometry (gate length for MOSFETS) and bias point. Both of these effects are shown in Figure 4.5 for bulk and SOI n-channel MOSFETS. There is certain drain current density that results in maximum cutoff frequency and it is almost constant for different technology nodes.

At a low frequency range (less than 15 GHz), the main amplifier branch in the Doherty structure is usually a deep class AB or class B biased amplifier. In higher frequencies, the small signal gain decreases considerably for class B biased amplifiers. For a class B biased amplifier, the low current density results in very low cutoff frequency, leading to low gain. However, the Doherty amplifier has a power splitter at the input. A portion of input power is delivered to the peaking amplifier, and at low input powers it is lost in the resistive part of the peaking amplifier's input impedance. In case of equal power division, the small signal gain decreases by 3 dB, according to the power division. Choosing class B bias for the main amplifier leads to very low small signal gain for the amplifier, unless there are driver stages in the main amplifier branch.

Maximum achievable output power is usually determined by the bias voltage and the size of the transistor. Scaling resulted in lower breakdown voltages in newer CMOS technology nodes. For a fixed output power, lower bias voltage means larger transistor size and higher bias current. The combination of low voltage and high current leads to low optimum impedances. This means that the design of matching networks for optimum impedance requires more complicated topologies and results in narrower operational bandwidth.

The size of the transistor is usually selected to have specific current density to achieve maximize gain or to achieve maximum linearity. As noted previously [18], an algorithm is given for selection of transistor size. This algorithm works acceptably for class A, AB, or B main amplifiers. For class C peaking amplifiers, larger transistors are required because the voltage and current of the transistor contain higher harmonics. For a class C biased transistor, maximum output power at fundamental frequency has the following relation with the conduction angle θ_c [19]:

$$P_{out} \propto \frac{\theta_c - \sin\theta_c}{1 - \cos(\theta_c/2)} \tag{4.9}$$

Because of reduced output power of the class C amplifier, larger transistors are usually needed for the peaking amplifier branch. As a rule of thumb, the size of the selected peaking transistor should be twice the size of the main transistor.

Output power and efficiency contours versus load impedance can be obtained using load-pull measurements or simulation. The load impedance at harmonic frequencies also affects the performance of the transistor. Figure 4.6 shows the effect of the second harmonic load impedance on efficiency value of a class AB biased common source 130-nm bulk CMOS transistor at peak input power. It can be seen that the value of efficiency can decrease by 8% if the second harmonic load is not considered in the design.

At mmW frequencies, passive lumped element parasitics have more deteriorating effects on the performance of the element. Parasitics also decrease the self-resonant frequency (SRF) of the lumped passive components. A 200-pH inductor in CMOS has a typical SRF between 50 and 130 GHz. As an example, when designing a 60-GHz amplifier with this inductor range, it is very difficult to control the load impedance at harmonic frequencies. Using extra elements for harmonic tuning also may not be practical because the losses from extra elements

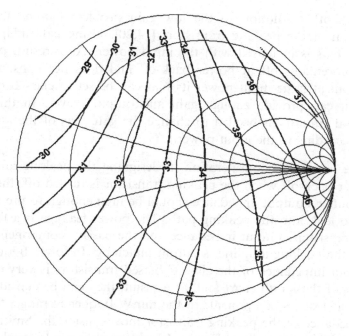

Figure 4.6 Effect of second harmonic load impedance on the efficiency of a 130-nm CMOS transistor at 60 GHz.

may reduce the efficiency more than the improvement expected from harmonic tuning. Transmission lines are a better choice if harmonic tuning is needed. By appropriately choosing the dimensions of the transmission lines, they are usable up to much higher frequencies than lumped elements. Using proper techniques, the losses associated with the transmission lines can be reduced. In the next section, the characteristics of passive elements are discussed in more detail.

The peaking amplifier is a class C biased amplifier that should be turned off at power back-off. A class C biased amplifier has very low maximum gain at mmW frequencies. This is mainly caused by two reasons. First, the harmonic content of the class C biased amplifier is high and the amount of fundamental output power is much lower than the class A or AB biased main amplifier. The second reason is that, to charge the input capacitance of the transistor, a higher amount of current is needed at higher frequencies. As a result, higher input power is needed to turn on the transistor at higher frequencies.

To have enough gain from the peaking amplifier, it should be implemented using more amplifier stages than the main amplifier's stages. Another solution to the low gain problem can be the gate adaptation of the peaking transistors [14,20]. Using gate adaptation, the gate bias is increased with the input power. As a result, peaking transistor remains off at power back-off and does not consume DC power, but at large input power its bias can be set to class B or class AB so it can provide enough gain and output power. In this case, additional circuitry is needed to change the gate bias of the amplifier stages according to the input power.

There is another consideration in designing the peaking amplifier. At power back-off when the peaking transistor is turned off, the peaking amplifier's output impedance should be high enough to prevent the power to leak into the peaking path. The power leakage is a function of the transistor's output impedance and the passive component losses used in the peaking amplifier's output matching. In lower frequencies, the output impedance of the class C biased transistors is very close to the edge of the Smith chart and, as a result, they can be considered as a (almost) lossless component. In the mmW frequency range, the output impedance of the peaking transistor moves inside the Smith chart, resulting in a higher loss. Figure 4.7 shows the small signal output impedance of a 130-nm NMOS transistor versus frequency. Low reflection coefficient impedance can be converted to low loss impedance using a proper matching network [21], but it should be noted that the matching network should be designed to present proper load impedance to the transistor itself. As a result, the lower reflection coefficient of the transistor usually leads to higher power leakage into the peaking branch in power back-off, resulting in lower efficiency.

4.3.2 Passive Components in mmW Doherty Structure

In Doherty amplifiers, passive components are used in impedance-matching, power-combining, and bias networks. Power loss associated with the passive components is the main limiting factor in designing mmW Doherty amplifiers. It is obvious that passive component losses decrease the maximum output power and efficiency of the amplifier. However, at power back-off these losses cause power loss into the peaking branch, resulting in further reduction of efficiency.

Other limiting factors in design are the parasitics and self-resonant frequencies of the passive components. SRF of the passive components can be on the order of second harmonic frequency. It makes harmonic

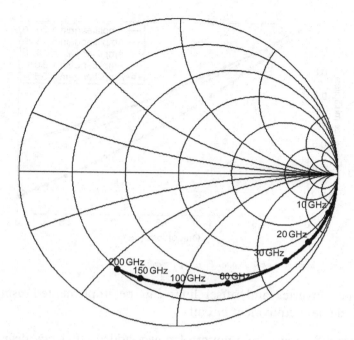

Figure 4.7 Small signal output impedance of a class C biased NMOS.

tuning of the transistors very difficult. As discussed in the previous section, having improper harmonic load impedance can decrease the efficiency considerably. Even in fundamental frequency, lumped element and layout parasitics should be taken into account in the design process because their effects in mmW frequencies can degrade the amplifier's performance considerably. Accurate models are needed for passive components, or electromagnetic simulation of passive components is needed in the design process.

Size is another factor that should be considered in the design. For analog circuits, the chip area used is usually defined by the size of the passive components. At a lower mmW frequency range, the wavelength in transmission lines is large. A typical quarter-wave transmission line with silicon dioxide as the dielectric is approximately 1.25 mm long at 30 GHz; however, by using slow-wave transmission lines, the physical length of the quarter-wave transmission lines can be reduced by a typical factor of 1.5–2.5. In some cases it may be impractical to implement the matching and combining networks using transmission lines at a lower mmW frequency range. In these cases, the

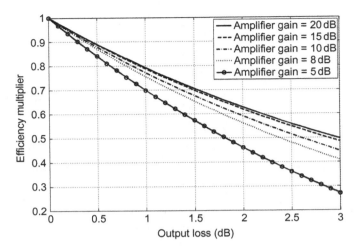

Figure 4.8 Effect of output network loss on the PAE of an amplifier.

impedance inverter and offset lines can be implemented using their lumped element equivalent circuits.

Figure 4.8 shows the amount of power-added efficiency degradation factor versus losses at the output of the amplifier for different amplifier gain values. A 1-dB loss at the output of an amplifier having 30% efficiency and 10-dB gain decreases the Power-Added Efficiency (PAE) to 22.5%. Output loss also decreases the maximum output power from the amplifier. In the Doherty amplifier, the output power losses are caused by the output matching networks, impedance inverters, offset lines, and power leakage into the peaking path. The matching network is present in other amplifier topologies. It has almost the same effect on the amplifier performance as the other amplifier structures. In this section, we discuss the other factors in the performance of the Doherty amplifier.

The impedance inverter and offset lines can be implemented using transmission lines when their lengths and form factors are not limiting issues. Because of a small separation of the metal layers in the back-end of the line, microstrip lines need very narrow signal lines for reasonable characteristic impedance. This increases resistive losses in microstrip transmission lines. The co-planar waveguide (CPW) transmission line is a more practical and popular structure in integrated designs. In CPW structure, the characteristic impedance can be controlled using the ground plane/signal line separation and the signal line width, resulting in a wide range of characteristic impedance choices.

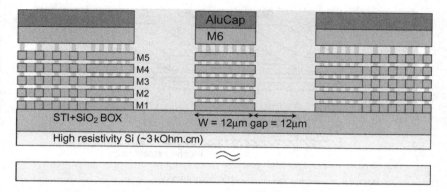

Figure 4.9 CPW transmission line utilizing multiple conductor layers [22]. © IEEE 2010.

The losses associated with the CPW transmission line are mainly caused by the resistive losses of the conductors and substrate loss and low skin depth in mmW frequencies. Conductor losses can be reduced by using thick conductor layers or by using multiple conductor layers, as shown in Figure 4.9 [22]. In CPW lines, a large portion of current flows on the edges of the conductors. Using thick conductors or multiple conductor layers increases the area that the current is flowing through, leading to lower resistive loss.

In CPW structure, electromagnetic fields can penetrate into the silicon substrate. For SOI substrates, because of high resistivity of substrate (approximately a few $k\Omega$.cm), it does not considerably increase the transmission line's loss. In bulk CMOS technologies substrate has low volume resistivity, resulting in induced currents in the lossy substrate and ultimately an increase in the transmission line's loss. Shielded CPW transmission lines are used to decrease the transmission line's loss. Figure 4.10 shows the structure of a shielded CPW transmission line. In this structure, the metal strips in the lower metal layers are used to prevent the fields from penetrating into the lossy substrate. The direction of the metal strips is perpendicular to the propagation direction of the transmission line. It does not allow the current flow in the metal strips, thus preventing conductive loss in the metal strips. However, the metal strips increase the parallel capacitance of the transmission line, resulting in a higher propagation constant (smaller wavelength). As a result, for a given electrical length, a shorter length of line is needed, resulting in size reduction and further reduction in transmission line loss.

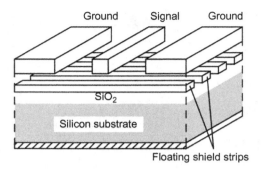

Figure 4.10 Shielded CPW transmission line structure [23]. © IEEE 2006.

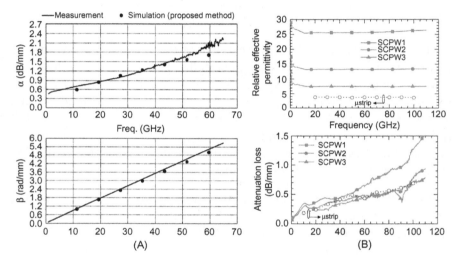

Figure 4.11 Loss and propagation constant for a shielded CPW transmission line (A) [24] (© IEEE 2009)
(B) [25] (© IEEE 2012).

The values of loss and propagation constant for a shielded CPW transmission line with characteristic impedance of approximately 44 Ω are shown in Figure 4.11. In this figure, a quarter-wave transmission line's loss at 60 GHz is in the range of 0.2−0.6 dB.

Power loss in a lossy network is also a function of load impedances connected to it. For the quarter-wave impedance inverter of the Doherty amplifier, the termination impedances are the load impedance and the output impedance of the main amplifier. The main amplifier is usually designed to have maximum output power at peak input power. As a result, its output reflection coefficient is not zero. Herein, for simplicity we assume that the main amplifier's output impedance it is matched to is R_{opt}. On the other side, the quarter-wave impedance inverter is terminated to R_L. Assuming that at peak power the output powers from the two amplifiers are equal, the value of R_L should be equal to $R_{opt}/2$.

The value of power loss in the quarter-wave transmission line can be obtained as:

$$P_{loss} = |a_1|^2 + |a_2|^2 - |b_1|^2 - |b_2|^2 = (1 + e^{-2\alpha l}(|\Gamma_L|^2 - |\Gamma_L|^2 e^{-2\alpha l} - 1))|a_1|^2$$

(4.10)

In Eq. (4.10), α is the real part of the complex propagation constant. For $R_L = R_{opt}/2$, $|\Gamma_L| = 1/3$ and the value of power attenuation can be calculated as:

$$P_{loss,dB} = -10 \log(e^{-2\alpha l}(e^{-2\alpha l} + 8)/9) = -10 \log\left(e^{\frac{-2Al}{8.69}}\left(e^{\frac{-2Al}{8.69}} + 8\right)/9\right)$$

(4.11)

In Eq. (4.11), A is the value of transmission line loss expressed in dB/mm. For the characteristics shown in Figure 4.11A, the value of attenuation at a frequency of 60 GHz is calculated as:

$$\alpha = 2.1 \text{ dB/mm}, \quad l = 0.31 \text{ mm}, \quad P_{loss,dB} = 0.72 \text{ dB}$$

(4.12)

Referring to Figure 4.8, this amount of loss decreases the efficiency of an amplifier with a 10-dB gain by a factor of 0.82.

The impedance inverter can be designed using a lumped equivalent circuit of the quarter-wave transmission line [26]. Lumped elements can lead to higher losses compared with the shielded CPW transmission line structures. Lumped inductors usually cannot be shielded from substrate because capacitive parasitics between inductor and shielding conductors decrease the resonant frequency of inductors such that they become useless at mmW frequency range. Figure 4.12 shows the losses of a 130-pH inductor with a quality factor of 47 at a frequency of 60 GHz versus the load impedance seen by the inductor. The other side of inductor is matched for calculating loss, as is the case in impedance-matching networks. The losses are shown for series and parallel configurations.

(A)   (B)

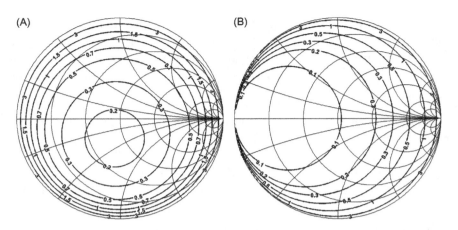

Figure 4.12 A 130-H inductor loss at 60 GHz versus load impedance seen by the inductor (A) series configuration and (B) parallel configuration.

The values of losses for a 53fF Metal-Insulator-Metal (MIM) capacitor and quality factor of more than 80 at 60 GHz are shown in Figure 4.13 for series and parallel configurations. The values of losses for a 53fF Metal-Oxide-Metal (MOM) capacitor and quality factor of more than 100 at 60 GHz are shown in Figure 4.14 for series and parallel configurations.

The simulated loss of a quarter-wave impedance inverter designed with the lumped element equivalent circuits for the characteristic impedance of 50 Ω is shown in Figure 4.15. As can be seen from this figure, the loss of the transformer with a MOM capacitor is in the same range as the quarter-wave transmission lines.

Passive component losses also increase the power loss into the peaking branch. The ratio of power leakage to the main amplifier's output power can be obtained as:

$$\frac{P_{\text{leak}}}{P_{\text{main}}} = \frac{\text{real}(Y_{\text{out},P})}{1/R_{\text{L}} + \text{real}(Y_{\text{out},P})} \quad (4.13)$$

In Eq. (4.13), $Y_{\text{out},P}$ is the output admittance of the peaking amplifier at power back-off. In addition to the peaking transistor's output impedance, the matching network elements cause additional losses in the peaking branch at power back-off, leading to lower efficiency at power back-off. As noted previously [21], using the two-sided matching

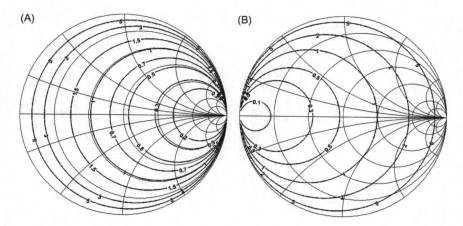

Figure 4.13 A 53fF MIM capacitor loss at 60 GHz versus load impedance seen by the capacitor (A) series configuration and (B) parallel configuration.

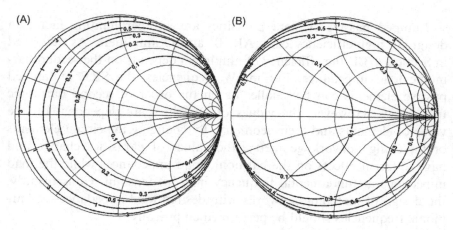

Figure 4.14 A 53fF MOM capacitor loss at 60 GHz versus load impedance seen by the capacitor (A) series configuration and (B) parallel configuration.

technique, the output impedance of the possible values of the peaking amplifier is obtained for a lossless matching network when the matching network is designed to obtain maximum output power at peak power. Also, it is shown how the network can be designed for low power leakage at power back-off while obtaining high output power using the reversed two-sided matching. For lossy matching network, however, the analysis is too complicated and depends on the quality factor of elements used and network topology, but the lossless approach can be used as a starting point for calculating power leakage into the peaking branch at power back-off.

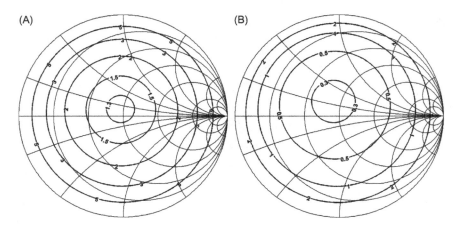

Figure 4.15 A lumped element 50-ohm quarter-wave impedance transformer's loss (A) using MIM capacitor and (B) using MOM capacitor.

Lumped passive element parasitics are another limiting factor in designing high-efficiency class AB, B, and C amplifiers. As described in Section 4.3.1, the efficiency of amplifiers depends on the second harmonic load impedance. For mmW frequencies, the SRF of lumped passive elements can be smaller or on the order of second harmonic frequencies. However, the values of parasitics are dependent on the value of elements and their geometry. By changing the circuit elements or matching network geometry, the values of parasitics change and have different effects on the harmonic load impedances while the load impedance at fundamental frequency may be constant. Consequently, the design of matching networks with desired load impedances at harmonic frequencies should be performed empirically.

Transmission lines are more robust in this point of view. For transmission lines, because of dispersion, the propagation constant changes with frequency. Also, the characteristic impedance has some changes with frequency; however, because these parameters are fixed or there is a transmission line with a predefined geometry, they can be considered for the second harmonic load-tuning and the design procedure can be performed more easily.

4.3.3 Other Techniques

Reducing the number of elements in the Doherty structure can further improve its performance. The impedance inverter and the offset lines add additional loss to the output, resulting in lower

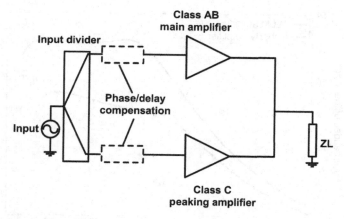

Figure 4.16 TLLM amplifier's structure.

efficiency for the amplifier. As noted previously [21], the transformer-less load-modulated (TLLM) architecture and its design procedure are presented for designing amplifiers with high efficiency at power back-off. The structure of the TLLM amplifier is shown schematically in Figure 4.16. Similar to the Doherty amplifier, the TLLM amplifier works based on the load modulation technique. It also uses a class AB (or class B) main amplifier and a class C peaking amplifier, but it does not utilize impedance inverters or the offset lines. For the main amplifier, the proper impedance transformation is done using the matching network; for the peaking amplifier, the low leakage condition is obtained using the matching network itself without the need for additional offset lines.

Figure 4.17 shows the simulated performance of a TLLM amplifier compared with the performance of a Doherty amplifier at 60 GHz with the same matching network topology. The combined loss of the impedance transformer and main branch's offset line in the Doherty amplifier is considered to be 0.7 dB, and the loss associated with the offset line in the peaking branch is considered to be 0.2 dB. Because of excessive loss of the impedance transformer and offset lines, maximum output power of the Doherty amplifier is lower than the TLLM amplifier by 0.8 dB. The excessive loss also causes lower efficiency for the Doherty amplifier. Because the passive component losses in the mmW frequency range are higher compared with the lower frequencies, the TLLM amplifier can be a candidate for high-efficiency mmW amplifier design.

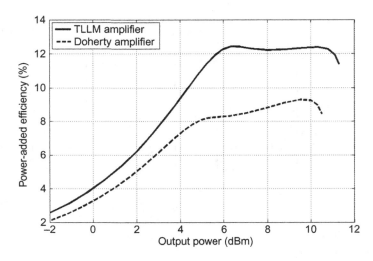

Figure 4.17 Simulated performance of 65-nm CMOS TLLM amplifier compared with the Doherty amplifier at 60 GHz.

REFERENCES

[1] Doherty WH. A new high efficiency power amplifier for modulated waves. Proc IRE 1936;24(9):1163−82.

[2] Darraji R, Ghannouchi FM, Hammi O. Generic load-pull-based design methodology for performance optimization of doherty amplifiers. IET Sci Meas Technol 2012;6(3):132−8.

[3] Yang Y, Cha J, Shin B, Kim B. A fully matched N-way Doherty amplifier with optimized linearity. IEEE Trans Microwave Theory Tech 2003;51(3):986−93.

[4] Kim B, Kim J, Kim I, Cha J. The Doherty power amplifier. IEEE Microwave Mag 2006;7 (5):42−50.

[5] Darraji R, Ghannouchi FM. High efficiency doherty amplifier combining digital adaptive power distribution and dynamic phase alignment. In: IEEE wireless and microwave technology conference; April 2013. p. 1−3.

[6] Campbell CF. A fully integrated Ku-Band Doherty amplifier MMIC. IEEE Microwave Guided Waves Lett 1999;9(3):114−16.

[7] Kobayashi KW, Oki AK, Gutierrez-Aitken A, Chin P, Yang L, Kaneshiro E, et al. An 18−21 GHz InP DHBT linear microwave Doherty amplifier. In: 2000 IEEE radio frequency integrated circuits (RFIC) symposium digest; June 10−13, 2000. p. 179, 182.

[8] McCarroll CP, Alley GD, Yates S, Matreci R. A 20 GHz Doherty power amplifier MMIC with high efficiency and low distortion designed for broadband digital communication systems. In: 2000 MTT-S international microwave symposium digest; June 11−16, 2000. p. 537−40.

[9] Tsai J, Huang T. A 38−46 GHz MMIC Doherty power amplifier using post-distortion linearization. IEEE Microwave Wireless Compon Lett 2007;17(5):388−90.

[10] Campbell CF, Tran K, Kao M, Nayak S. A K-Band 5W Doherty amplifier MMIC utilizing 0.15 μm GaN on SiC HEMT technology. In: 2012 Compound semiconductor integrated circuit symposium (CSICS); October 14−17, 2012. p. 1−4.

[11] Wang X, Peng Y, Ma F, Sui W. A Ka-band MMIC Doherty power amplifier using GaAs pHEMT technology. In: 13th international symposium on integrated circuits (ISIC); December 12–14, 2011. p. 91–3.

[12] Taghian F, Abdipour A, Mohammadi A, Roodaki PM. Nonlinear analysis and design of a mm-wave wideband Doherty power amplifier. In: 2011 international conference on electrical and control engineering (ICECE); September 16–18, 2011. p. 6003–6.

[13] Wicks B, Skafidas E, Evans R. A 60-GHz fully-integrated Doherty power amplifier based on 0.13-μm CMOS process. In: 2008 IEEE radio frequency integrated circuits (RFIC) symposium digest; June 2008. p. 69–72.

[14] Agah A, Dabag H, Hanifi B, Asbeck PM, Buckwalter JF, Larson LE. Active millimeter-wave phase-shift Doherty power amplifier in 45-nm SOI CMOS. IEEE J Solid-State Circuits 2013;48(10):2338–50.

[15] Shopov S, Amaya RE, Rogers JWM, Plett C. Adapting the Doherty amplifier for millimetre-wave CMOS applications. In: 2011 IEEE 9th new circuits and systems (NEWCAS) international conference; June 26–29, 2011. p. 229–32.

[16] Niknejad AM, Hashemi H. mm-Wave silicon technology 60 GHz and beyond. New York, USA: Springer; 2008.

[17] Dickson TO, Yau KHK, Chalvatzis T, Mangan AM, Laskin E, Beerkens R, et al. The invariance of characteristic current densities in nanoscale MOSFETs and its impact on algorithmic design methodologies and design porting of Si(Ge) (Bi)CMOS high-speed building blocks. IEEE J Solid-State Circuits 2006;41(8):1830–45.

[18] Yao T, Gordon MQ, Tang KKW, Yau KHK, Yang M, Schvan P, et al. Algorithmic design of CMOS LNAs and PAs for 60-GHz Radio. IEEE J Solid-State Circuits 2007;42 (5):1044–57.

[19] Kraus HL, Bostian CW, Raab FH. Solid state radio engineering. New York, USA: John Wiley; 1980.

[20] Cripps SC. Advanced techniques in RF power amplifier design. Norwood, USA: Artech House; 2002.

[21] Akbarpour M, Helaoui M, Ghannouchi FM. A transformer-less load-modulated (TLLM) architecture for efficient wideband power amplifiers. IEEE Trans Microwave Theory Tech 2012;60(9):2863–74.

[22] Siligaris A, Hamada Y, Mounet C, Raynaud C, Martineau B, Deparis N, et al. A 60 GHz power amplifier with 14.5 dBm saturation power and 25% peak PAE in CMOS 65 nm SOI. IEEE J Solid-State Circuits 2010;45(7).

[23] Cheung TD, Long JR. Shielded passive devices for silicon-based monolithic microwave and millimeter-wave integrated circuits. IEEE J Solid-State Circuits 2006;41(5).

[24] Vecchi F, Repossi M, Eyssa W, Arcioni P, Svelto F. Design of low-loss transmission lines in scaled CMOS by accurate electromagnetic simulations. IEEE J Solid-State Circuits 2009;44(9).

[25] Franc A, Pistono E, Gloria D, Ferrari P. High performance shielded coplanar waveguides for the design of CMOS 60-GHz bandpass filters. IEEE Trans Microwave Theory Tech 2011;59(5):1219–26.

[26] Kang D, Kim D, Cho Y, Park B, Kim J, Kim B. Design of bandwidth-enhanced Doherty power amplifiers for handset applications. IEEE Trans Microwave Theory Tech 2011;59 (12):3474–83.

CHAPTER 5

Reliable Power Amplifier

Baudouin Martineau
STMicroelectronics, Crolles, France; Univ. Grenoble Alpes, F-38000 Grenoble, France; CEA, LETI, MINATEC Campus, F-38054 Grenoble, France

5.1 INTRODUCTION

The integrated CMOS power amplifier is somewhat recent [1]. The main challenge was to achieve the RF performances with a device dedicated to digital. This has been possible because of the increase of f_T/f_{max} together with gate length reduction. The evolution toward deep nanometer CMOS technologies (65 nm and below) has facilitated the design of highly integrated systems for consumer applications. Today, CMOS PA design has been demonstrated up to millimeter wave, and most connectivity products integrate the PA in the transceiver. Nevertheless, even if the scale reduction of the CMOS offers the opportunity to address higher and higher frequencies, the counterpart is reliability issues. For PA design, the use of nanometer CMOS technologies brings major challenges, some of which had not been encountered before. Such challenges include thermal effects, metal interconnects electromigration (EM), time-dependent dielectric breakdown (TDDB), hot carrier injection (HCI), and electrostatic discharge (ESD) events. Power amplifier reliability is a task not only for technology development but also for designers.

This chapter discusses the different reliability constraints that occur in deep CMOS technologies. Some reliability-aware CMOS PA design examples and methodologies are also presented.

5.2 EFFECT OF CMOS TECHNOLOGY SCALING ON THERMAL MANAGEMENT

The reliability of silicon devices is frequently estimated by accelerated tests performed at high temperatures to generate failures in a short time period. For instance, burn-in method is often used as a reliability screen to weed out infant mortality. This method is used because most failures

Linearization and Efficiency Enhancement Techniques for Silicon Power Amplifiers.
DOI: http://dx.doi.org/10.1016/B978-0-12-418678-1.00005-2

are accelerated because of elevated temperature. Accordingly, as a starting point, the temperature of any power amplifier must be known to estimate its reliability. Heat is produced in both the substrate and the interconnections. The main source of heat generation is the power dissipation of devices that are embedded in the substrate. Some power dissipation also results from Joule heating caused by the flow of current in the interconnects, but this is mostly negligible when regarding the total power dissipated. There are three ways to dissipate temperature: infrared radiation, convection, and conduction. Radiation can be considered negligible because of the very small area of an integrated circuit. Convection of a circuit itself is also very small for the same reason. However, if a heatsink is added, then the convection becomes the main way to extract the temperature, and thus the thermal resistance can be drastically reduced. Nevertheless, in most consumer connectivity applications, this type of solution cannot be applied because of the small packaging volume available. In such a case, the conduction becomes the predominant mechanism to dissipate temperature. Equation (5.1) shows how the junction temperature of the transistor can be evaluated.

$$T_j = T_a + P \times R_{th} \tag{5.1}$$

where T_a is the ambient temperature and P (in W) is the total power consumption including the DC and the RF dissipated in the device. For RF, the RMS power value is usually a good approximation. R_{th} is the overall thermal resistance that represents the sum of different contributions, as depicted in Eq. (5.2).

$$R_{th} = R_{cb} + \left(R_{jsi} + \frac{R_{jp}}{\text{gnd\_con}} // R_{jc} \right) // R_{heatsink} \tag{5.2}$$

The diverse thermal resistances are named as follows:

- R_{cb} is the case-to-board resistance
- R_{jsi} is the junction resistance
- R_{jp} is the junction to pad resistance
- R_{jc} is the junction to case resistance
- $R_{heatsink}$ is the heatsink resistance (in case of use)

gnd_con represents the number of contacts connected to the ground pads.

Figure 5.1 gives examples of thermal resistance.

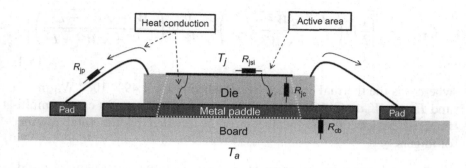

Figure 5.1 Chip on-board thermal resistance representation.

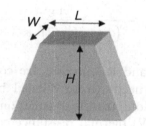

Figure 5.2 Truncate pyramid shape illustration.

Thermal resistance of R_{cb} and R_{jc} propagate like a truncate pyramid into the materiel. This pyramid has a diffusion angle of approximately $40°$. The truncate pyramid shape thermal resistance can be calculated with Eq. (5.3):

$$R_{th\_tpy} = \frac{H}{((L + H.\tan(40°)) \times (W + H.\tan(40°))) \times \lambda} \qquad (5.3)$$

where λ is the thermal conductivity of the materiel, H is the height of the truncate pyramid, and L and W are the length and the width of the top part, respectively. These geometrical parameters are illustrated in Figure 5.2.

Usually R_{jsi} is negligible, but in nano CMOS this resistance becomes significant. Recently, a relationship between the thermal resistance of a MOSFET and its geometrical parameters was derived using 3-D heat flow. Equation (5.4) is obtained for bulk technologies based on work performed previously [2].

$$R_{jsi} = \frac{1}{2\pi\lambda} \left[\frac{1}{L} \ln\left(\frac{L + \sqrt{W^2 + L^2}}{-L + \sqrt{W^2 + L^2}} \right) + \frac{1}{W} \ln\left(\frac{W + \sqrt{W^2 + L^2}}{-W + \sqrt{W^2 + L^2}} \right) \right]$$

(5.4)

where λ is the thermal conductivity of silicon ($\lambda = 1.49 \times 10^{-4}$ W/μm °C) and L and W are CMOS channel geometry parameters. In case of multi-fingered devices, the R_{jsi} value must be divided by the total number of fingers.

R_{jp} represents any junction to pad connections, such as wire bonding or solder bump. It can be well-estimated by using the cylinder thermal resistance Eq. (5.5):

$$R_{jp} = \frac{length}{(diameter/2)^2 \times \lambda \times \pi}$$

(5.5)

Assuming that DC and RF ground are connected to the thermal ground, the R_{jp} value needs to be divided by the total number of ground connections. Note that this equation can also be used for through-silicon via (TSV) thermal resistance calculation. Some typical values for bump and bonding can be found in Table 5.1.

Table 5.2 gives the thermal conductivity of different materials used in microelectronics.

In conclusion, a good starting value for packaging thermal resistance can be estimated to be between 15 and 60°C/W, depending on the packaging and connection used. A remarkable point is that mm-wave packages often use ceramic material together with nanometer scale CMOS transistor circuits. This combination is very damaging for good thermal resistance. In addition, mm-wave CMOS PA usually has low efficiency. In that case, a careful thermal study must be performed to predict reliability issues.

Table 5.1 Typical R_{th} Values for Various Connection Types		
Connection type	R_{th}	Comments
Solder bump	300°/W	Pad width ~60 μm
Copper pillar bump	70°/W	Pad width ~40 μm
BGA ball	75°/W	
Gold wire bonding	1600°/W	1 mm length; 50 μm cross-section

Table 5.2 Thermal Conductivity for Different Microelectronic Materials	
Material	Thermal conductivity (W/cm °C)
Ag	4.18
Cu	3.9
Al	2.37
Pb	0.35
Sn	0.666
Or	3.17
Si	1.49
SiO$_2$	0.014
Ni	0.4
Epoxy	0.0025
Air	0.000262

5.3 METAL INTERCONNECTS ELECTROMIGRATION

Metal wire and contact can tolerate only a certain amount of current density. When this current density is too high, an EM effect appears. Electromigration is the displacement of metal atoms in a semiconductor. It happens when the current density is high enough to cause the drift of metal ions in the direction of the electron flow. The effect is important in applications when high direct current densities are used, such as with power amplifiers. Deep CMOS process technology is increasing the impact of EM and the problem of the voltage drop on the supply rails. This drop is $\Delta V = IR$, where I is the current and R is the supply rail resistance. This is often called IR-drop effect. These two phenomena have an impact on the performance and reliability of the power amplifier. Driven by Moore's law, metal interconnect widths are decreasing exponentially. Thus, the overall cross-sectional area of interconnect is shrinking. In addition, currents are not scaling proportionally to shrinking wire widths; therefore, modern transceivers have extremely high current densities.

Advanced CMOS process technologies have very complex rules for current density limits that are dependent not only on wire widths but also on wire lengths, as well as on the dimensions of the wire layers above and below a via layer. Current limits in metallization wiring are proposed to ensure reliable operations over a given time period without

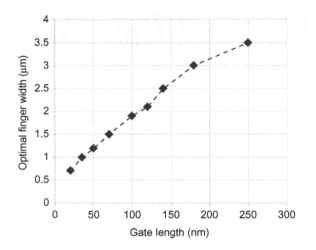

Figure 5.3 Simulated f_{max} optimal finger width for various NMOS minimum technology gate lengths.

significant EM damage. Usually, process technology current limits are created by considering 0.1% failure over 10 years for a given maximum current density in a copper cross-section at maximum qualified process temperature. In power amplifier design, current limits are usually restrictive for chock inductor and other components that drive the DC to the transistor. In that case, the EM reliability constraints are often not compatible with desired device characteristics and/or metal density design rules. In addition, nanometer CMOS transistors have an optimum sizing antagonistic to EM rules. Figure 5.3 shows the f_{max} optimal finger width with respect to various technology gate lengths.

It appears that in nanoscale CMOS (<100 nm), the best performance is achieved for minimum finger width. However, at such a small finger size, only a few contacts can be placed to drive the current. In those conditions, some current density values cannot be sustained in reliable condition. A solution commonly used at RF frequencies consists of using two contact rows instead of one, as illustrated in Figure 5.4. Unfortunately, this solution significantly reduces the f_{max} and is not adapted to millimeter wave designs.

5.4 TIME-DEPENDENT DIELECTRIC BREAKDOWN (TDDB)

TDDB and hot carrier injection (HCI) are the two main causes of failure in the NMOS transistor. For TDDB, caution must be taken

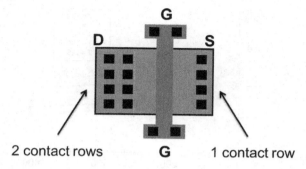

2 contact rows   G   1 contact row

Figure 5.4 NMOS with one and two contact rows at drain and source, respectively.

regarding the maximum RMS gate voltage, which is equivalent to electric field strength across the gate oxide. Each electric stress consumes the oxide lifetime. This means that this degradation is cumulative. To protect against TDDB induced by a combination of current, voltage, and temperature, maximum gate voltage guidelines (V_{max} values) as a function of the junction temperature are usually given by fab process rating. The time to failure can be calculated using Eq. (5.6)

$$ttf = f(V, \text{Area}, T) \tag{5.6}$$

Advanced CMOS node (<65 nm) has a gate oxide thinner and a lower operating voltage. Hence, the initial rupture in the oxide does not significantly disrupt the functionality of the transistor. Therefore, to determine the process fab rating, the device parameters are monitored after the oxide breakdown. The failure time is determined by the time it takes for the transistor performance to degrade to a certain level. This criterion is used to meet the qualification guidelines. However, there is a possibility that, for power amplifiers, even the initial break can cause issues with the functionality of the particular part. When the part initially breaks, although the transistor performance is not significantly degraded, there is a large increase in the amount of noise for the gate leakage current.

As an example, the values in Figure 5.5 are fab guidelines to insure reliable performance of a 65-nm CMOS single-oxide NMOS with a cumulative failure of 0.1% per 100,000 power-on device hours or >10 years.

When an RF swing voltage is applied to a power amplifier, the TDDB reliability check can be predicted by performing a time domain

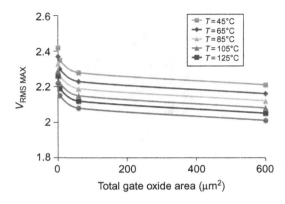

Figure 5.5 Fab guidelines example to insure reliable performance of a CMOS 65-nm single-oxide NMOS for >10 years.

voltage simulation on transistor nodes (V_{dg}, V_{gs}, and V_{gb}) at the most extreme dynamic operating points (i.e., saturated output power in a power amplifier). This should confirm that the voltage RMS waveform exposure of the gate does not exceed the maximum allowed voltage.

5.5 HOT CARRIER INJECTION

HCI degradation in FET device characteristics occurs when the drain-to-source bias exposure is high over time. Because the transistor channel is starting at low V_{ds}, the electric field is equally divided over the length of the channel by the existence of an inverted channel. When the drain−source voltage is increased, the channel begins to pinch-off, restraining the electric field to the region between the channel edge and the drain diffusion. This causes an increase of the magnitude of the electric field and channel carriers to accelerate through the pinched-off region, reaching a velocity that is greater than the thermally limited diffusion drift velocity. If a channel "hot carrier" strikes a crystal atom near the drain region, then it may produce an electron−hole pair. In that case, the transistor transconductance slowly degrades. Figure 5.6 illustrates the mechanism.

This mechanism is usually monitored by the degradation of the I_{dsat} over time. Thus, to insure a given lifetime for a specific power amplifier, the DC and RF voltage must be examined. RF lifetime must be evaluated considering the application, the mission profile, and the modulation scheme used. These two last points are discussed in the next sections.

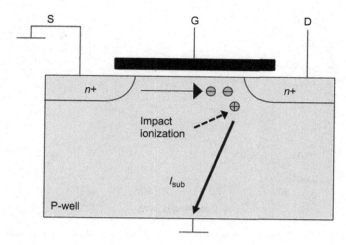

Figure 5.6 HCI in nMOSFET.

5.5.1 DC Model

HCI effects limit the voltages that may be used with CMOS devices and are most severe at shorter channel lengths. The HCI device degradations are correlated with the substrate current for nMOSFETs. The correlation exists because hot carriers are driven by the maximum channel electric field, which occurs at the drain end of the channel. By monitoring the substrate current, it is possible to evaluate the degradation. Detailed models to calculate device degradation during circuit operation and to simulate the impact on circuit operation have been presented [3] and are synthesized in Eq. (5.7).

$$I_{sub}(V_{gs}, V_{ds}) = \frac{Ai}{Bi}(V_{ds} - V_{dsat}(V_{gs}))I_{ds}(V_{gs})e^{-Bi.lc/(V_{ds}-V_{dsat}(V_{gs}))} \quad (5.7)$$

$$lc = (lc_0 + lc_1 V_{ds})\sqrt{t_{ox}} \quad (5.8)$$

$$V_{dsat} = E_0 L_{eff}(V_{gs} - V_t)/(E_0 L_{eff} + (V_{gs} - V_t)) \quad (5.9)$$

Ai, Bi, lc_0, and lc_1 are technology-dependent parameters. E_0 is the critical electrical field and L_{eff} is the effective channel length. For the calculation of device lifetime, the degradation of a parameter is written as:

$$\Delta D = \left(\frac{I_{ds}}{H.W}\left(\frac{I_{sub}}{I_{ds}}\right)^m t\right)^n \quad (5.10)$$

where H, m, and n are technology-dependent parameters, t is the stress time, and W is the channel width.

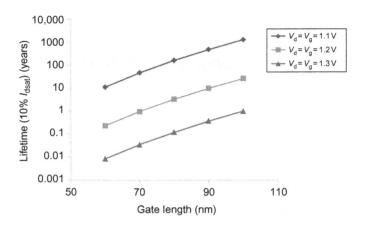

Figure 5.7 Simulated DC lifetime with respect to the nMOSFET channel length for various V_{ds} voltage.

Figure 5.7 shows the simulated DC lifetime (10% I_{dsat} degradation criterion) with respect to the nMOSFET channel length for various V_{ds} voltages. Simulations have been performed in 65-nm CMOS technology with single-oxide devices in worst case conditions ($T = 125°C$, $V_{ds} = V_{gs}$).

This simulation shows that a small reduction in V_{ds} can enormously improve the lifetime of the device. Most of the time, if an optimum f_T/f_{max} is targeted, then it is better to slightly decrease the V_{ds} instead of increasing the gate length to reach the desired lifetime.

5.5.2 RF Model
Recent CMOS power amplifier publications [4−6] have estimated HCI reliability with a commonly used reliability model [7]. In that case, lifetime is defined as the time of the drain current of the output stage of the PA to decrease by 10%. In case of multi-stage PA, the last stage is chosen to monitor the degradation because it is subject to a higher RF voltage stress in comparison with previous stages. The lifetime is assumed to follow the exponential relationship given in Eq. (5.11).

$$\tau = c_1 e^{\left(c_2 / V_{ds peak}\right)} \tag{5.11}$$

where c_1 and c_2 are empirical parameters that must be extracted for each V_{ds} and V_{gs} and temperature conditions.

Another modeling method is now proposed by some technology manufacturers. In that case, the design kit includes a MOSFET aging

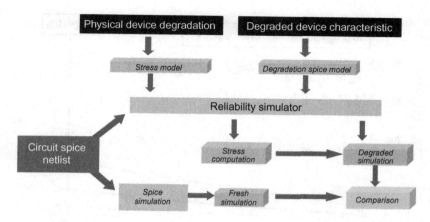

Figure 5.8 Typical reliability simulation flows.

model. The reliability analysis consists of at least two runs, one for assessment (fresh run) and another one for performing degraded simulation (degraded run). In the fresh run, the stress environment is examined at each instant of simulation time and recorded independently for each of the devices in the circuit. Note that this first run must be a transient run to provide the behavior of the device during time. At the end of this run, the damages that are linked to the stimulus provided (this includes the voltages, currents, duration of the stress, as well as the temperature) are extrapolated to the time that is specified in the aging-related commands. Figure 5.8 shows the reliability simulation flow available in STMicroelectronics 65-nm CMOS technology.

Figure 5.9 shows NMOS I/V curves. In addition, bias points for 10% I_{dsat} degradation criterion have been included. These points have been extracted with the simulation flow depicted in Figure 5.8. They include DC and DC + RF points with respect to various V_{ds}/V_{gs} voltages and RF Vswing. Simulations have been performed in 65-nm CMOS technology with single-oxide devices.

This simulation demonstrates that RF stress must be taken into account when the biasing point is chosen. In addition, this simulation shows that HCI NMOS reliability is also conditioned by the class of operation. Because the gate-to-source voltage is lower, a class B or C case is a more favorable than a class A case.

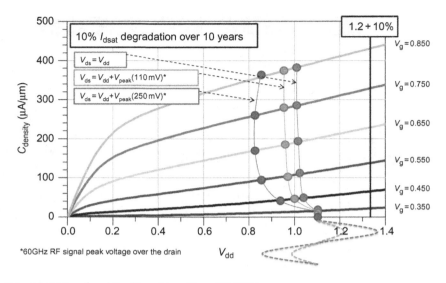

Figure 5.9 10% I_{dsat} degradation bias points on 65-nm CMOS I/V curves.

5.6 ELECTROSTATIC STATIC DISCHARGE

There are three types of ESD events that can destroy a CMOS circuit and, thus, a CMOS power amplifier. These are HBM (human body model), MM (machine model), and CDM (charged device model). Figure 5.10 illustrates these three use cases.

5.6.1 Human Body Model

HBM is used to emulate the transfer of electrostatic charge through a series resistor from the human body to a device. When a human moves across the ground, an electrostatic charge accumulates on the body. In that case, simple contact of a part of the body (a finger, for instance) with the pins of a device or assembly allows the body to discharge, possibly causing device destruction. The model used to simulate this event is the HBM. The HBM testing model represents the discharge from the fingertip of a human delivered to the device. This model is represented in Figure 5.11.

For HBM simulation of a power amplifier, this model is added together with the complete PA schematic including ESD diodes or other protections. Then, the simulation is performed in a time domain and each part of the circuit is probed to insure that no peak voltage can damage sensitive components such as transistors.

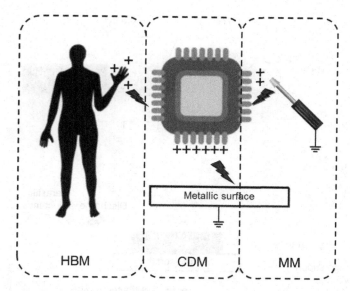

Figure 5.10 HBM, CDM, and MM case illustrations.

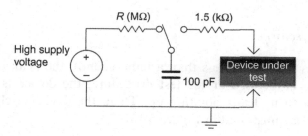

Figure 5.11 HBM circuit.

5.6.2 Machine Model

MM event is linked to a charged conductive object, such as a metallic utensil or charged cables. This ESD model consists of a 200-pF capacitor discharged directly into a component with no series resistor. The 0.5 pH inductance used in the schematic Figure 5.12 represents the parasitic element that shapes the oscillating machine model wave form.

5.6.3 Charged Device Model

CDM occurs when different parts of the circuit discharge via one pin, creating a large voltage difference across the chip. Once again, this event can be destructive if there is no ESD strategy. This event can be more critical than the HBM for some devices. Even if the length of the

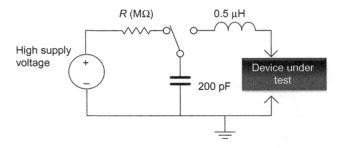

Figure 5.12 MM circuit.

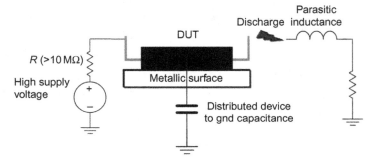

Figure 5.13 Typical CDM test circuit.

discharge is very short (less than a nanosecond), the peak current can reach some amperes. For the test procedure, the device is placed on a surface with its leads pointing up. Then, the device is charged and discharged, as illustrated in Figure 5.13.

5.6.4 ESD Protection

ESD protection strategies in RF input/output (I/O) are made by large area diodes and/or specific clamp transistors. All I/O such as RF signal, digital bits, ground, and V_{dd} are connected to ESD-protected pads as illustrated in Figure 5.14. In addition, small diodes in the reverse-bias direction are usually distributed in a long metal connection inside the circuit. Nevertheless, these solutions are not compatible with mm-wave signal. Parasitic capacitance of the PN junction of ESD diodes represents low impedance at mm-wave frequency and thus shortens the I/O signal.

Some solutions to bypass this issue have been found. Most of them have been proposed for input low-noise amplifier (LNA) protection, but they can be easily adapted to a power amplifier. Figure 5.15 illustrates the solution developed previously [8].

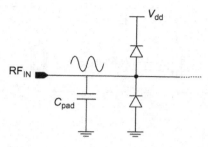

Figure 5.14 RF input with ESD diodes protection.

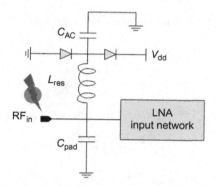

Figure 5.15 RF input with ESD diode protection connected after a parallel inductance.

Similarly, connecting multiple parallel stubs to insure the ESD protection has been proposed [9]. Figure 5.16 illustrates this strategy.

This second solution offers better protection because the hypothetical ESD event is distributed and, thus, the maximum voltage protection is improved. However, the input or output power-matching can be more difficult to achieve and less efficient. Another option has been proposed [10]. In that case, the author took advantage of the output balun to isolate the power amplifier devices from the ESD event. Figure 5.17 shows this approach.

5.7 VOLTAGE STANDING WAVE RATIO

RF power amplifiers usually have an output impedance of 50 Ω and, in an ideal world, would only be connected to load impedance of 50 Ω. In that case, 100% of the power is absorbed in the load and 0% of the

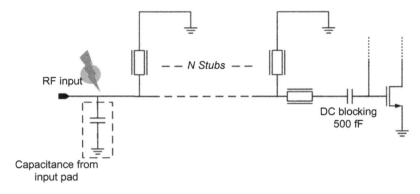

Figure 5.16 RF input with ESD protection insured by multiple parallel stubs.

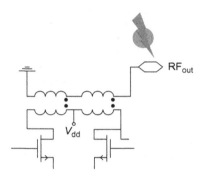

Figure 5.17 RF output with ESD protection insured by the secondary of a balun.

power is reflected back to the power amplifier. Nevertheless, in the real world, RF amplifiers are used together with devices that load impedances that are approximately 50 Ω. In addition, these load impedances vary with frequency. A reliable power amplifier must be capable of dealing with load mismatches without degrading performance or lifetime.

Voltage standing wave ratio (VSWR) is the ratio of the amplifier output (source) to the connected device input impedance (load) at a given frequency. If the amplifier is connected on a perfect load impedance of 50 Ω, then the VSWR is equal to 1:1. A relationship between the impedance Z, gamma, VSWR, and the mismatch losses (ML) can be written as described in Eq. (5.12).

$$Z = Z_0 \cdot \text{VSWR} \quad \text{with} \ Z_0 \ \text{the reference impedance}$$

$$\Gamma = \frac{\text{VSWR} - 1}{\text{VSWR} + 1}$$

$$S_{\text{returnlosses}}(\text{dB}) = 20 \cdot \log(\Gamma)$$

$$ML(\text{dB}) = 10 \cdot \log(1 - \Gamma^2)$$

(5.12)

Table 5.3 synthesizes some values of VSWR.

One solution to avoid mismatch problems can be to use broadband-matching networks to match the output impedance of the amplifier to the load. Unfortunately, the output impedance of the power amplifier along with the load impedance varies as a function of frequency. In addition, all devices like cables and connectors contribute to mismatch because they are not an ideal 50 Ω across a wide frequency range. A short or open, even if they are quick, constitutes a VSWR $= \infty$. In that case, all the power is reflected back to the power amplifier. Equation (5.13) gives the reflected power in W and dBm with respect to the VSWR.

$$P_{\text{ref}} = P_{\text{in}} \left(\frac{\text{VSWR}-1}{\text{VSWR}+1} \right)^2 \text{with } P \text{ in W}$$

$$P_{\text{ref}}(\text{dBm}) = 10 \cdot \log \left(10^{\frac{P_{\text{in}}}{10}} \left(\frac{\text{VSWR}-1}{\text{VSWR}+1} \right)^2 \right) \text{with } P \text{ in dBm}$$

(5.13)

VSWR	Return loss (dB)	Mismatch loss (dB)	Gamma	Z (load) high (W)	Gamma	Z (load) low (W)
1	μ	0.000	0	50.00	0	50.00
2	9.542	−0.512	0.333	100.00	−0.333	25.00
3	−6.021	−1.249	0.5	150.00	−0.5	16.67
4	−4.437	−1.938	0.6	200.00	−0.6	12.50
5	−3.522	−2.553	0.667	250.00	−0.667	10.00
6	−2.923	−3.100	0.714	300.00	−0.714	8.33
7	−2.499	−3.590	0.75	350.00	−0.75	7.14
8	−2.183	−4.033	0.778	400.00	−0.778	6.25
9	−1.938	−4.437	0.8	450.00	−0.8	5.56
10	−1.743	−4.807	0.818	500.00	−0.818	5.00

Table 5.3 RF Output with ESD Protection Insured by the Secondary of a Balun

A reliable power amplifier must be capable of absorbing reflected power from extreme mismatches encountered in normal conditions. To evaluate the over-voltage swing generated by the reflected power, Eq. (5.14) can be used.

$$
\begin{aligned}
V_{\text{swing}} &= \sqrt{Z_0 \cdot P_{\text{out}} \left(\frac{\text{VSWR}-1}{\text{VSWR}+1} \right)^2} \cdot \sqrt{2} \text{ with } P \text{ in } W \\
V_{\text{swing}} &= \sqrt{Z_0 \cdot 0.001 \cdot 10^{(10 \cdot \log(10^{(P_{\text{in}}/10)}((\text{VSWR}-1)/(\text{VSWR}+1))^2)/10)}} \\
&\quad \cdot \sqrt{2} \text{ with } P \text{ in dBm}
\end{aligned}
\tag{5.14}
$$

Amplifiers that are unable to carry large reflected power require protection. One issue could be when using a class AB amplifier. The transistor in a class AB amplifier is biased to produce output current for less than 360° and more than 180° of the input signal. A class AB design consumes less power in its quiescent state than when an input signal is applied. In class AB, the improvement in efficiency in comparison with that of the class A allows use of smaller transistors and the silicon chips used can be smaller, resulting in less heat and a smaller electrical path to insure EM rules.

Some protection techniques can be used to safeguard the last stages from reflected power. One approach is simply attaching an attenuator between the output of the power amplifier and the load. By doing so, the bad load of the VSWR is enhanced and the reflected power is as well-attenuated. With this solution, there is less reflected power and any reflected power is decreased by n dB by the attenuator. The weakness is that the forward power into the load is also attenuated by n dB. This is why this solution cannot be used in low-power or mm-wave applications. Another option consists of real-time monitoring of the transistor junction temperature of the power amplifier. When the temperature goes above a predetermined value, the amplifier immediately shuts down or the power is reduced. A parallel approach could be to directly screen the reflected power and, when a maximum threshold is hit, to shut-down the amplifier [11]. This technique can be improved by monitoring the reflected power and change the gain of the power amplifier or decrease the driver output power level as the reflected power increases. This method is often called "foldback" and is used to insure that the reflected power never exceeds the maximum admissible level. Nevertheless, any of these techniques will protect the amplifier to some extreme cases, such as a defective cable or load shorts or opens, resulting in an infinite VSWR. As a result, 100% of the forward power is reflected back into the output stages of the power amplifier. This incident happens more often than one may think.

5.8 POWER AMPLIFIER DESIGN FOR RELIABILITY

The power amplifier design for reliability can be achieved following the process described in Figure 5.18.

5.8.1 Mission Profile Analysis

The mission profile of a product is the timeline of the environmental conditioning and usage of the product. To define the mission profile, several questions must be answered regarding the following:

- Target product lifetime with respect to the target application (connectivity, cellular, etc.) and market (consumer, professional, medical, military, etc.)
- Product used in standby mode (hours/day); Tx active time (% of standby)
- Product used in power-on mode (hours/day); Tx active time (% of power-on)

A mission profile for a connectivity application is given in Table 5.4.

The mission profile analysis must be the first step in the reliability analysis because it provides the power amplifier lifetime requirement, which can be very different than the one of the product. In the example given in Table 5.4 regarding 10-year product lifetime, the final Tx (i.e., the PA) lifetime decreases to 1826 days (5 years). Another aspect of the mission profile must be analyzed: the modulation scheme used. High data rate communication requires complex modulation such as OFDM (orthogonal frequency division multi-plexing). For instance, the signal waveforms used in IEEE 802.15.3c WirelessHD standard have, in the time domain, very large ($\sim$ 10 dB) peak-to-average spikes. An OFDM HRP2 (16QAM) modulation signal is shown in Figure 5.19. In practical use, the PA typically operates at a 10-dB back-off from the 1-dB compression point (denoted mean power of 0 dB in Figure 5.19).

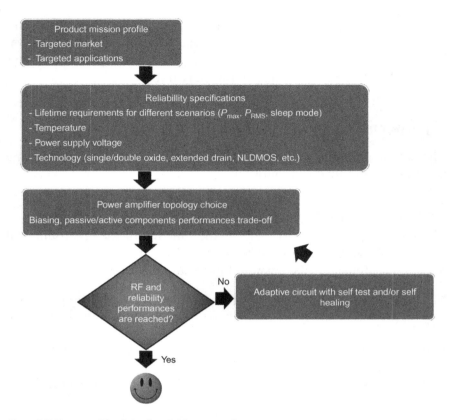

Figure 5.18 Power amplifier design for reliability process flow.

Table 5.4 Mission Profile for TX Lifetime Requirement		
Product mission profile		
The mission profile of a product is the timeline of the environmental conditioning and usage of the product		
Product lifetime	10 years	
Product use in standby mode	12 hours/day	43,800 hours over lifetime
TX active time (% of standby)	0.02%	9 hours
Product use in connect on mode	12 hours/day	43,800 hours over lifetime
TX active time (% of power on)	100.00%	43,801 hours over lifetime
TX active standby + connect	43,810 hours or	1826 days over 10 years (43,824 hours/5.00 years)
RX active standby + connect	87,600 hours or	3651 days over 10 years (87,624 hours/10.00 years)
Requirement for the stressed devices with respect to the expected life of the application (TX)	1 times	
Final TX lifetime requirement	43,810 hours or	1826.00 days over 10 years (43,824 hours/5.00 years)

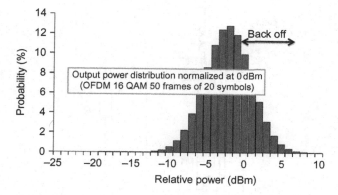

Figure 5.19 WirelessHD™ standard HRP2 mode signal distribution.

Table 5.5 Power Amplifier Lifetime Requirement According to Tx Lifetime Condition and the Modulated Power Distribution		
Tx lifetime requirement	43,810 hours (∼5 years)	
Lifetime distribution (% of total):	PA @ mean power	76.5%
	PA @ P1dB	22.5%
	PA @ maximum power	1%
Lifetime distribution (hours/years):	PA @ mean power	33,515 hours 3.83 years
	PA @ P1dB	9857 hours 1.12 years
	PA @ maximum power	438 hours 0.05 years

Considering the Tx mission profile together with the modulation used, the power amplifier lifetime requirement can be split. Table 5.5 synthesizes the example given in Table 5.4 and the signal distribution of Figure 5.19. It can be noticed that, most of the time, the PA is not at the maximum stress level (i.e., maximum output power). Finally, the 10-year product lifetime can be achieved if the power amplifier is able to support 3.83 years at mean power, 1.13 years at output compression point, and 0.05 years at maximum output power.

5.9 INTRINSICALLY ROBUST DESIGN

Reliability-aware CMOS power amplifier design examples can be found in the literature. Most of them are focused on the splitting of the output voltage swing. To do that, one of the most common techniques is the cascode topology [12,13]. The basic idea is to share the total

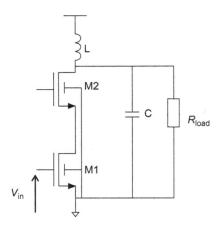

Figure 5.20 Simplified schematic of a cascode power amplifier.

output swing between the two NMOS. Figure 5.20 provides an illustration of the cascode structure.

It can be noticed that the common gate (M2) device can be different than the common source device (M1). A good strategy can be to use a high-voltage-capable MOSFET for M2 and a high-performance MOSFET for M1. It is preferable to use the high-speed transistor as a common source device because it is the main contributor of the maximum available RF gain of the cascode core. In that configuration, the output swing is not equally shared between the two transistors. On the contrary, the Vds voltage values can be 1 V and 2.5 V for the single and double oxide devices respectively. Some solutions have also been reported with extended drains or NLDMOS devices for the common gate device.

One could imagine that this solution can be used with a number of MOSFET stacked to sustain any voltages. Unfortunately, there is a limitation because of the biasing of the P-substrate. As explained previously, it is possible to share the voltage between the different drains, gates, and sources, but the last transistor of the stack sees a high drain-to-substrate voltage. To overcome such issues, two strategies can be used. The first one has been described elsewhere [14] and is illustrated in Figure 5.21. In standard cascode configuration, a high RF peak voltage at the drain of the last device can lead to a current flow through the drain diode and to a clipping of the output voltage waveform. At higher RF peak voltages, the drain diode can be destroyed. To avoid breakdown of the drain diode, the last device is realized in a

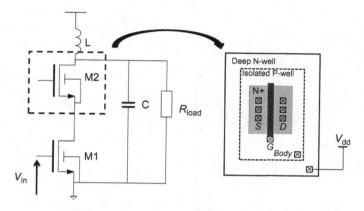

Figure 5.21 Cascode power amplifier with isolated common gate transistor.

p-well inside a deep n-well, with the n-well biased to the supply voltage. The connection to the source of the common gate forces the p-well to swing together with the top cascode device, reducing the voltage across the drain diode.

The second strategy was proposed previously [15] and revisited recently [16]. The circuit is composed of a common source input transistor and three common gate stacked transistors connected in series, as shown in Figure 5.22. This topology typically benefits from the use of silicon in insulator substrate technology (SOI). Because of the insulator, each transistor is "floating," meaning that there is no biasing of the P-substrate, therefore eliminating the drain diode breakdown issue. In this "totem"-like topology, the output swings are added in phase. Unlike in a cascode configuration, where the gate of the common gate transistor is RF-grounded, a small external gate capacitance (C2, C3, and C4) is added to allow an RF swing at the gate of each stacked transistor. This guarantees that each transistor has the same drain−source, drain−gate, and gate−source voltage swings and the absolute voltage swings increase with respect to ground. A good point is that the method is also compatible with advanced CMOS processes such as 28-nm and 14-nm FDSOI.

Splitting the output voltage swing is not the only strategy to achieve an intrinsically robust design. Another way can be to use a small voltage and summing power or current to achieve the desired output power. Figure 5.23 shows the magnitude of the peak voltage swing seen at the transistor level with respect to the output power. In this

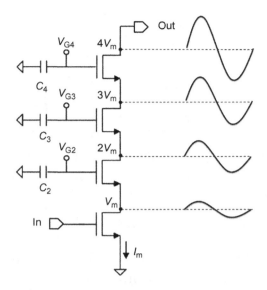

Figure 5.22 Stacked CMOS floating body power amplifier structure.

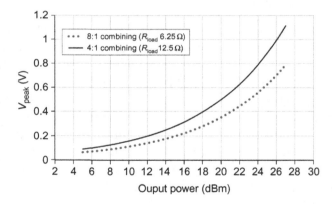

Figure 5.23 Voltage peak value as a function of the combining ratio and the output power (including 1-dB combiner loss).

graphic, eight-to-one and four-to-one combining use cases have been evaluated. In both cases, 1-dB losses have been taken into account for the power combiner. This technique is particularly well-adapted at millimeter wave frequency when deep submicron CMOS is used [10].

5.10 SELF-HEALING DESIGN

Self-healing is a method to eliminate reliability performance reduction over time (such as HCI effect) by identifying any degradation and

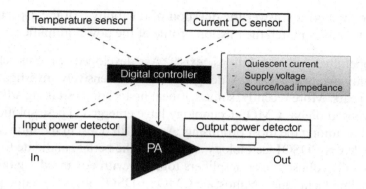

Figure 5.24 Self-healing power amplifier illustration.

modifying the circuit to restore its performance after fabrication by incorporating a feedback loop into the system. In addition, this type of solution takes advantage of the digital processing available in CMOS processes. To adjust the performance of the circuit after fabrication, some elements must be implemented together with the power amplifier core. Usually, a power detector is used to evaluate the output power level [17]. This power detector is associated with an ADC and a digital controller. It can be possible to enlarge the self-healing capabilities by adding some other sensors, as proposed previously [18]. Among them, DC current and temperature sensors are the most interesting. One example of the healing flow chart is given Figure 5.24. In this example, the controller receives the data provided by the calibrated power sensor at the PA output and returns different voltage signals to the PA. These voltages can be used to change the quiescent current, to switch on or off some parallel paths, to adjust the source or load impedance, or to modify the power supply value. The goal is to adapt the shape gain profile of the PA to overcome process variations and gain compression to meet the linearity specifications.

5.11 CONCLUSION

This chapter has shown that the use of nanometer CMOS technologies brings major challenges for PA design. The reliability aspect in CMOS power amplifier design must be taken into account early in the feasibility study. Packaging thermal resistance, technology option, modulation scheme, and targeted market have a huge impact on reliability constraint together with "classical" power amplifier figure of merit. There is not a generic method to answer the question of reliability. All the aspects must be dealt with to match the mission profile of the power amplifier.

Hopefully, some solutions exist. As developed in this chapter, reliability-aware CMOS PA design has been intensively investigated for many years. More recently, some self-healing techniques taking advantage of advanced digital CMOS technology have emerged. These solutions are probably among the most promising. Against all odds, in the future, nanoscale CMOS FDSOI technology is probably the better candidate to design intrinsically robust power amplifiers together with advanced digital-based self-healing techniques. Nanoscale CMOS FDSOI offers the same properties as the first SOI technology generation [15,16]. Because of the insulator, each transistor body can be controlled independently and therefore eliminates the drain diode breakdown issue. In addition, the high level of integration of such technology offers an advantage to integrate numerous sensors together with ADC and digital controllers.

Finally, I say:

Reliability is, after all, engineering in its most practical form
James R. Schlesinger, Former US Secretary of Defense [19].

REFERENCES

[1] Chang JY-C, Abidi AA. A 750 MHz RF amplifier in 2-/spl mu/m CMOS. In: Symposium on VLSI circuits, 1993. Digest of Technical Papers, May 19−21; 1993. p. 111−2.

[2] Semenov O, Vassighi A, Sachdev M, Keshavarzi A, Hawkins CF. Effect of CMOS technology scaling on thermal management during burn-in. IEEE Trans Semiconductor Manuf 2003;16(4):686−95.

[3] Hu C, Tam SC, Hsu F-C, Ko PK, Chan TY, Terrill KW. Hot-electron induced MOSFET degradation—model, monitor and improvement. IEEE Trans Electron Devices 1985;ED-32:375−85.

[4] Kjellberg T, Abbasi M, Ferndahl M, de Graauw A, v.d.Heijden E, Zirath H. A compact cascode power amplifier in 45-nm CMOS for 60-GHz wireless systems. In: Compound semiconductor integrated circuit symposium, 2009 (CISC'09), annual IEEE, October 11−14; 2009. p. 1−4.

[5] Stephens D, Vanhoucke T, Donkers JJTM. RF reliability of short channel NMOS devices. In: Radio frequency integrated circuits symposium (RFIC'09), IEEE, June 7−9; 2009. p. 343−6.

[6] Zhao D, Kulkarni S, Reynaert P. A 60 GHz dual-mode power amplifier with 17.4 dBm output power and 29.3% PAE in 40-nm CMOS. In: Proceedings of the ESSCIRC 2012, September 17−21; 2012. p. 337−40.

[7] Guarin FJ, La Rosa G, Yang ZJ, Rauch SE, III. A practical approach for the accurate liftime estimation of device degradation I deep sub-micron CMOS technologies. In: IEEE international conference on devices, circuits and systems; April 2002.

[8] Siligaris A, Richard O, Martineau B, Mounet C, Chaix F, Ferragut R, et al. A 65-nm CMOS fully integrated transceiver module for 60-GHz wireless HD applications. IEEE J Solid-State Circuits 2011;46(12):3005–17.

[9] Marcu C, Chowdhury D, Thakkar C, Park J-D, Kong L-K, Tabesh M, et al. A 90 nm CMOS low-power 60 GHz transceiver with integrated baseband circuitry. IEEE J Solid-State Circuits 2009;44(12):3434–47.

[10] Martineau B, Knopik V, Siligaris A, Gianesello F, Belot D. A 53-to-68 GHz 18 dBm power amplifier with an 8-way combiner in standard 65 nm CMOS. In: Solid-state circuits conference digest of technical papers (ISSCC), 2010 IEEE International, February 7–11; 2010. p. 428–9.

[11] Gorisse J, Cathelin A, Kaiser A, Kerherve E. A 60 GHz 65 nm CMOS RMS power detector for antenna impedance mismatch detection. In: Proceedings of ESSCIRC'09, September 14–18; 2009. p. 172–5.

[12] Knopik V, Martineau B, Belot D. 20 dBm CMOS class AB power amplifier design for low cost 2 GHz–2.45 GHz consumer applications in a 0.13 μm technology. In: International symposium on circuits and systems (ISCAS), 2005 IEEE, May 23–26; 2005. p. 2675–8.

[13] Ruberto M, Degani O, Wail S, Tendler A, Fridman A, Goltman G. A reliability-aware RF power amplifier design for CMOS radio chip integration. In: Reliability physics symposium, April 27–May 1 (IRPS'08). IEEE International; 2008. p. 536–40.

[14] Leuschner S, Pinarello S, Hodel U, Mueller JE, Klar H. A 31-dBm, high ruggedness power amplifier in 65-nm standard CMOS with high-efficiency stacked-cascode stages. In: Radio frequency integrated circuits symposium (RFIC), 2010 IEEE, May 23–25; 2010. p. 395–8.

[15] Pornpromlikit S, Jeong J, Presti CD, Scuderi A, Asbeck PM. A 33-dBm 1.9-GHz silicon-on-insulator CMOS stacked-FET power amplifier. In: Microwave symposium digest, 2009 (MTT'09); June 2009.

[16] Chakrabarti A, Krishnaswamy H. High power, high efficiency stacked mmwave class-E-like power amplifiers in 45 nm SOI CMOS. In: Custom integrated circuits conference (CICC), 2012 IEEE, September 9–12; 2012. p. 1–4.

[17] Liu JY-C, Tang A, Ning-Yi W, Gu QJ, Berenguer R, Hsieh-Hung H, et al. A V-band self-healing power amplifier with adaptive feedback bias control in 65 nm CMOS. In: Radio frequency integrated circuits symposium (RFIC), 2011 IEEE, June 5–7; 2011. p. 1–4.

[18] Bowers SM, Sengupta K, Hajimiri A. A fully-integrated self-healing power amplifier. In: Radio frequency integrated circuits symposium (RFIC), 2012 IEEE, June 17-19; 2012. p. 221–4.

[19] O'Connor PDT. Practical reliability engineering. 4th ed. New York, NY: John Wiley & Sons; 2002ISBN 978-0-4708-4462-5.

CHAPTER 6

Efficiency Enhancement for THz Power Amplifier

Ullrich R. Pfeiffer and Neelanjan Sarmah
High-Frequency and Communication Technology, University of Wuppertal, Wuppertal, Germany

6.1 INTRODUCTION

Historically, terahertz radiation was generated by optical means with the help of femto-second laser pulses and was primarily used for spectroscopic purposes in the scientific community [1]. This band, which is defined by the sub-mmWave band from 300 GHz through 3 THz, is often referred to as the *THz gap*, loosely describing the lack of adequate technologies to effectively bridge this transition region between microwaves and optics. The use of either electronic or photonic techniques has led to the subdivison in *THz electronic* and *THz photonic* bands with frequencies spanning from 100 GHz through 10 THz, as illustrated in Figure 6.1.

At lower mm-wave frequencies at which silicon devices still exhibit power gain, amplifiers suffer from device limitations, such as their low breakdown voltages and reduced output impedance and high implementaton losses of passive matching networks. High-performance SiGe HBTs exhibit breakdown voltages of $BV_{ceo} = 1.5$ V and $BV_{cbo} = 4.5$ V [DOTSEVEN], whereas RF CMOS suffers from low drain voltages of approximately 1 V in 65-nm technologies [2].

At lower mm-wave frequencies, state-of-the-art SiGe power amplifiers without power-combining have demonstrated saturated output powers as high as 20 dBm at 60 GHz [3,4], up to 10 dBm at 160 GHz [5], and up to 5 dBm at 240 GHz [6]. Sub-mmWave amplifiers, however, need to operate transistors close to, or even above, their cutoff frequencies. Because of this, silicon-based sub-mmWave sources have only demonstrated comparatively low output power in the past. Power generation techniques beyond the cutoff frequency are performed in one of

Linearization and Efficiency Enhancement Techniques for Silicon Power Amplifiers.
DOI: http://dx.doi.org/10.1016/B978-0-12-418678-1.00006-4

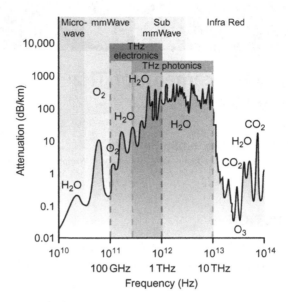

Figure 6.1 Atmospheric absorption showing the terahertz spectrum and the subdivision into THz electronic *and* THz photonic *bands indicating the different approaches for signal generation and detection.*

two ways: directly extracted from oscillators or up-converted from lower frequencies by the help of frequency multiplier chains.

Over the past years, tremendous progress has been made in the field of solid-state frequency multipliers. With the introduction of III/V MMICs, single-chip multiplier-chain solutions up to 300 GHz and 195 GHz have been demonstrated with an output power of −6 and 0 dBm, respectively [42]. Recent advances in silicon-based devices have also made them useful for multiplier-chain applications. Output powers of −3 dBm at 325 GHz and −29 dBm at 825 GHz have been reported [7,8] for a 0.13-μm SiGe HBT process. This monolithic multiplier is implemented in an f_{max}-optimized evaluation of SiGe HBT technology [9] using cascaded transistor-based multiplication and amplification stages. This result compares quite well with III/V-based mHEMT multiplier chains published previously [10] and demonstrates the capability of silicon technologies for power generation up to 825 GHz. A traveling-wave doubler implemented in a 65-nm CMOS process with an output power as high as −6.6 dBm at approximately 244 GHz has proven the viability of this technology for sub-mmWave applications [11]. The major drawbacks of today's existing sub-mmWave multiplier-chain solutions are very high DC power consumption and very low chip surface efficiency.

State-of-the-art fundamental frequency oscillators with the highest operation frequency of 573 GHz and an output power of −19 dBm have been reported for an InP DHBT device with f_{max} of 859 GHz [43]. Fundamental oscillators in silicon at such high frequencies are still unavailable. However, by applying novel design methodologies to both fundamental and N-push topologies, some new designs approximately 200 GHz and beyond have already been demonstrated for both SiGe HBT and CMOS technologies [44]. In particular, the work of [45] has reported an output power of −4.5 dBm for a SiGe HBT push−push Colpitts oscillator at 190 GHz. A fundamental frequency VCO in SiGe HBT technology covering the 218- to 248-GHz frequency range with up to −3.6 dBm output power and a combined VCO-doubler showing −1.7 dBm output power at 290 GHz have been demonstrated recently [12]. The output power achieved for CMOS designs is typically very small for many practical applications. Good examples are −47 dBm at 410 GHz in a 45-nm node, −46 dBm at 324 GHz, and −36.6 dBm at 553 GHz [13−15]. A novel and systematic approach applied for design of CMOS N-push oscillators allowed achievement of an output power of −17 dBm at 256 GHz and −7.9 dBm at 482 GHz for the three-stage single-end ring topologies in 130-nm and a 65-nm CMOS process nodes, respectively [16]. The very recent record output power levels in CMOS of −1.2 and −7.2 dBm measured on-wafer and radiated from large on-chip antenna arrays, respectively, have been reported [17,46].

Over the past decade, however, silicon has evolved as a formidable low-cost option with high f_{max}, as high as 500 GHz [18] and with 700 GHz in development [DOTSEVEN] [41]. Higher cutoff frequencies can be leveraged in two ways. A faster technology node can be used to bias devices at lower currents to achieve a given f_{max} performance. This leads to reduced power dissipation. However, faster technology can drive higher operating (e.g., carrier) frequencies with typically approximately one-third of peak f_{max} in practical applications, taking into account the various implementation losses. Hence, power generation in silicon technologies more than 300 GHz inevitably require multiplier chains. A typical implementation involves designing amplifiers with high output power up to $f_{max}/3$, which drives frequency multiplier chains with high input drive for strong harmonic generation. However, in general, it is seen that the saturated output power (P_{sat}) and the power-added efficiency (PAE) for PAs decrease with frequency.

This is due to the fact that at resonance, the real part of the output impedance goes down with frequency and makes it challenging to couple power from the transistor to the load. This negates the overall benefit one can have in terms of higher maximum available gain (MAG) from faster processes at frequencies more than 100 GHz. The reduction of output power also limits the overall PAE for power amplifiers.

In this chapter, we try to address the challenges involved in the design of PAs with high output power at frequencies approaching mm-wave and sub-mmWave band. The chapter is organized into four sections. Section 6.2 presents the power amplifier performance trade-offs toward THz operation with an emphasis on the conflicting design paradigms. Section 6.3 presents the device scaling considerations for future high-power THz, with a discussion of the limitations arising from process technology. A comprehensive analysis of this has already been presented [5]. Section 6.4 summarizes the state-of-the-art PA in silicon and III/V technologies.

6.2 POWER AMPLIFIER PERFORMANCE TRADE-OFFS TOWARD THZ OPERATION

Power amplifier performance trade-offs can be expressed by several key parameters. This includes the output power P_{out}, power gain G, operating frequency f, linearity in terms of the third-order input intercept point (IIP3), and PAE. The following Eq. (6.1) shows a figure of merit for power amplifiers (FoM_{PA}) derived from the system drivers section of the ITRS document [19]. The FoM_{PA} is limited to linear PAs for ease of comparison. To the first order, the one-pole system response of a transistor current or voltage gain exhibits a 20-dB per decade roll-off over frequency. The FoM_{PA} therefore includes a factor $(f/f_{max})^2$ to compensate the power gain roll-off. The operation frequency f is normalized with the unity power gain f_{max} to make the FoM_{PA} independent of the technology capabilities.

$$FoM_{PA} = P_{out} \cdot G \cdot PAE \cdot \left(\frac{f}{f_{max}}\right)^2. \qquad (6.1)$$

The optimal technology choice for high-power applications in the mm-wave regime is driven by two factors: the technology should be fast enough to provide sufficient power gain at the desired operation frequency and the output power level should be high enough to meet the system requirements. The following describes various design trade-offs between the key parameters of Eq. (6.1).

6.2.1 Output Power Limitations in Silicon Process Technologies

The maximum power P_{out} of Eq. (6.1) that an amplifier can deliver to an external load R_{load} (e.g., 50 Ω) depends, to the first order, on three technology factors: (i) the breakdown voltage (BV) of the device; (ii) the maximum collector current densities (j_c); and (iii) the achievable impedance transformation ratio (r). This simply reflects $P_{out} = (BV)^2/rR_{load}$, where $r = R_{load}/R_{in}$ is the impedance transformation ratio of the amplifier output-matching network. The impedance R_{in} is provided by the matching network and is seen at the device level, for example, at the collector in a common emitter transistor configuration. R_{in} defines the load-line of the amplifier. In other words, the achievable output power of a single device is independent of the operating (e.g., carrier) frequency. Although, at higher frequencies the amplifier's gain may be too low to provide sufficient amplification. High-performance silicon technologies therefore provide various device performance options to trade-off output power (e.g., breakdown voltages) versus cutoff frequencies.

Figure 6.2 shows the scaling of the breakdown voltage (BV_{ceo}) for high-performance SiGe HBT technologies. The data, courtesy of IBM [20], shows the breakdown voltage versus the peak cutoff frequency f_T for various technology nodes. As the SiGe technologies advance to faster cutoff frequencies, their breakdown voltages initially decline rather quickly and then level-off to approximately 1 V. This BV_{ceo} roll-off is a fundamental device limit first explained by Johnson in 1965 [21]. The Johnson limit assumes that, to the first order, BV_{ceo} can be calculated from the product of a critical electrical field E_{crit} at which carrier multiplication by impact ionization runs away, times the width of the collector base space charge region W_{CB}. If one reduces W_{CB} to increase f_T (shorter transit time), then one automatically reduces BV_{ceo}. According to the Johnson limit, the following rule-of-thumb applies:

$$f_T \cdot BV_{ceo} \leq 200 \text{ GHzV} \qquad (6.2)$$

The data in Figure 6.2 initially follows the Johnson limit very nicely; however, advanced technologies show some deviation from the simple Johnson limit. Note, the BV data are shown versus f_T, the short

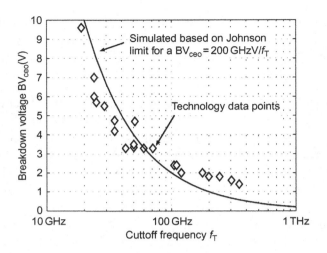

Figure 6.2 Breakdown voltage versus peak cutoff frequency (f_T) for various SiGe process technologies nodes. Data courtesy of IBM [20].

circuit current gain, instead of the more relevant unity power gain f_{max} because f_T is generally available in literature and less prone to extraction and measurement errors. In general, f_T and f_{max} are correlated.

The maximum available power in a given technology node can be estimated from the maximum voltage swing across the transistor's load-line impedance R_{in}. The voltage swing is limited by impact ionization, which can be overcome to a certain extent if the bias circuit provides a low enough external base resistance [22,23]. Low external resistance provides an escape path for hot carriers and, hence, works against the avalanche process taking place in a HBT device. As a result, the base current will change its polarity for the duration of the ac-swing that is within the avalanche region. In practice, BV_{ce} can be approximately 50% higher than BV_{ceo} as long as BV_{cbo} is not exceeded. For a class A operation, the maximum available power can be estimated as:

$$P_{out} = \left(\frac{x BV_{ceo} - V_{knee}}{2\sqrt{2}}\right) / R_{in} \qquad (6.3)$$

where $V_{knee} \approx 0.3$ V is the knee voltage at the transistor's saturation region, BV_{ceo} is the open base breakdown voltage, $R_{in} = R_{load}/r$ is the load-line impedance provided by the output-matching network, and x represents a 50% BV_{ceo} increase due to the bias network impedance. Based on Eq. (6.3), one can estimate the maximum available output

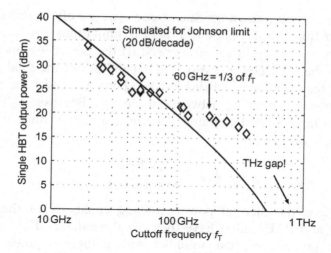

Figure 6.3 *Maximum single-device output power trend versus cutoff frequency (f_T). The data are shown for single transistors operating in a class A mode without power-combining. External base resistance supports a 50% extension of BV_{ceo}, a load-line impedance (R_{in}) of 10 Ω with $r = 5$, and $R_{load} = 50$ Ω. Note that other values in Eq. (6.3) may shift these data within 3 dB or more, but its trend remains the same.*

power for the BV_{ceo} data shown in Figure 6.2. The output power, however, strongly depends on the achievable load-line impedance. At high frequencies, low R_{in} is difficult to achieve due to design limitations in the output-matching network. For instance, high-quality factors for resonant matching networks in silicon are only approximately 10−20, making it difficult to achieve an impedance transformation r much larger than 5 in realistic applications. Under such conditions one can calculate the single device output power trend of Eq. (6.3), which is shown in Figure 6.3. The data suggest that at 60 GHz, for example, one-third of $f_T = 180$ GHz, single transistor amplifiers should be able to reach output powers as high as 20 dBm. These observations are supported by recent publications in this frequency range.

The data in Figure 6.3 initially follow the Johnson limit, with a 20-dB/decade roll-off; however, as one gets closer to approximately 1 THz, the output power exhibits roll-off faster than $1/f_T$. This is known as the THz gap, at which xBV_{ceo} in Eq. (6.3) equals the knee voltage V_{knee} and creates zero output power. At that point the breakdown voltage is too small to make the transistor switch—a fundamental limit for electronics systems trying to bridge the THz gap.

Power-combining techniques are often used to enhance the total output power of an amplifier by combining the power of several in parallel operating transistors. Such techniques are often combined with an impedance transformation. Practical limitations for such power combiners are set by three parameters: (i) the power loss in the combiner; (ii) the total power dissipation capabilities of the final amplifier and package assembly; and (iii) by the ability to model the parasitics of large power-combining networks accurately. Because of this, power combiners have limited gain and efficiency, which, in turn, affect their output power as follows:

$$P_{out} = \frac{P_{dc} \cdot PAE}{(1 - (1/G))} \qquad (6.4)$$

where P_{dc} is the thermal power-handling capability of the package assembly and PAE and G have their usual meanings. For instance, a packaged power-combined amplifier with a thermal power-handling capability of 4 W and a total power gain of 6 dB at a peak PAE of 10% can only generate up to 530 mW per chip. These considerations show that higher output powers can only be achieved if the individual transistor amplifiers are able to operate at high efficiency and high power gain.

6.2.2 Power Gain Trade-offs

The favorite design technique for high-frequency power amplifiers is to cascade many gain stages to form one circuit with high overall power gain. Because each amplifying device also contributes its own high frequency roll-off, the last stage in the amplifier chain should compress first for optimum linearity. The design of the last amplifier stage is most critical in terms of overall amplifier performance and is the focus here. The gain of the last stage depends on the circuit architecture, the intrinsic device gain, and the electrical losses in the input- and output-matching networks. A low overall gain mainly affects the amplifier efficiency, whereas a high gain stage exhibits larger negative feedback and tends to be more unstable.

The loss of impedance-matching networks is most critical at the output. A 3-dB loss, for instance, affects the efficiency differently at the input or output. With α and β being the loss of the input- and output-matching network, respectively, the peak power-added efficiency (PAE$_{peak}$) is:

$$PAE_{peak} = \left(\frac{1}{\beta} - \frac{\alpha}{G}\right)\eta \qquad (6.5)$$

where $\eta = P_{out}/P_{dc}$ is the amplifier's drain efficiency. Equation (6.5) shows that a higher power gain G can reduce the absolute power lost at the input, whereas loss accrued at the output may not be compensated. For instance, the PAE_{peak} roll-off for a $G = 16$-dB power amplifier with a drain efficiency of $\eta = 10\%$ is shown in Figure 6.4A. In case of loss-less matching, the PAE_{peak} versus amplifier gain is shown in Figure 6.4B. As a result, high gain stages like cascode amplifiers are preferred over common emitter stages.

6.2.3 PAE Trade-offs

Applications in the mm-wave frequency range may require a transistor bias point at approximately one-third of the technology peak f_T/f_{max}. This imposes a limit on their achievable PAEs because high-performance SiGe devices inherently have a steep f_T roll-off at high injection. The roll-off is due to the high germanium content and a large germanium gradient in the neutral base that enables the high beta and f_T of the device [24]. At peak efficiency, the dc bias current through the device is increased up to a point beyond the peak of the f_T curve and the device happens to be biased at a collector current density that heavily compresses the amplifier gain and therefore limits the available efficiency for class A operation to only a few percent [25]. Higher efficiency modes of operation are difficult to achieve. Early results have shown peak efficiencies up to approximately 20.9% [26], although their low power gain (4.2 dB) makes these amplifiers impractical for realistic applications. Recently, PAEs of 29.2% have been reported for PA at 60 GHz with a gain of 13.5 dB [27]. This suggests that high PAEs can only be achievable if the device is operated at

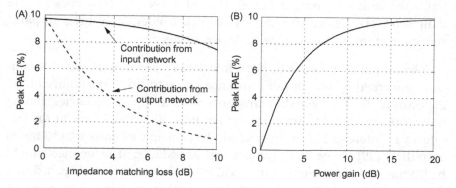

Figure 6.4 (A) PAE_{peak} roll-off for a 16-dB power amplifier with a $\eta = 10\%$ drain efficiency. (B) PAE_{peak} versus G for loss-less matching.

sufficient back-off, for example, approximately one-tenth of its peak f_T/f_{max}. This can be seen by re-writing the standard PAE equation [28] in terms of output power as shown in Eq. (6.6).

$$PAE = \frac{P_{out} - P_{in}}{P_{DC}} = \frac{P_{out}(1 - (1/\text{Gain}))}{P_{DC}} \quad (6.6)$$

$$\frac{dPAE}{dP_{out}} \alpha \frac{1}{\text{Gain}} \text{ and } \frac{dPAE}{d\text{Gain}} \alpha \frac{1}{\text{Gain}^2} \quad (6.7)$$

From Eq. (6.7), it is evident that PAE is a strong function of gain. For advanced process technologies like DOTFIVE, the MAG for a cascode amplifier is 5−6 dB at 250 GHz in the ideal case. However, taking into account the implementation losses, it is typically 2−3 dB per stage. Hence, gain-peaking techniques whereby some amount of instability is deliberately introduced can be an option to enhance the gain, for example, adding an inductance at the base of the common base transistor [29].

6.2.4 Impedance Match for Maximum Power Delivery
The load-line approach for designing PA requires finding the optimum impedance R_{opt} at the transistor output, which simultaneously maximizes the large signal voltage and current swing. In general, for a process technology, the optimum load resistance R_{opt} for a cascode amplifier is given by $R_{opt} = (BV_{CBO} - V_{knee})/(2I_{DC} - I_{knee})$, where BV_{CBO} is the collector base breakdown voltage and V_{knee} is the knee voltage. This has been discussed in detail previously [28]. Here, I_{DC} is the dc bias current, and this corresponds to current density for peak f_{max}. Hence, larger device sizes at the output stage are required to maximize the saturated output power P_{sat}. However, at frequencies more than 100 GHz, device size cannot be scaled indefinitely because parasitic capacitances also scale with device sizing.

The output of a transistor has a low pass response, with the time constant governed by the output resistance and parasitic capacitance at the collector node. To make the transistors work at high frequencies, tuning inductors are added to tune-out the net parasitic capacitance at the frequency of interest. This is, in general, the principle of narrowband tuned amplifiers [29]. In general, the net real impedance at resonance should be frequency-independent in the ideal case. However, in reality, it shows frequency dependence and goes down with increasing frequency. A reduced output resistance makes it challenging to couple power from

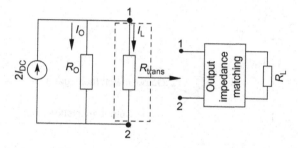

Figure 6.5 Current source equivalent of the output stage of the PA. R_O is the output impedance of the cascode amplifier and R_{trans} is the transformed load impedance R_L [5].

the transistor to the load. This can be explained by simple current source modeling of the output stage, as shown in Figure 6.5. If R_O is low compared with R_{trans}, then most of the power is dissipated in the transistor than what is coupled to the load.

6.2.4.1 Load-Line Match Versus Conjugate-Matching

Based on the current source modeling of the output stage, the value of the transformed load impedance R_{trans} is given by Eq. (6.8). Here, R_{opt} is the desired optimum load impedance (from load-line analysis) and R_O is the output resistance of the transistor.

$$R_{trans} = \frac{R_O R_{opt}}{R_O - R_{opt}} \tag{6.8}$$

The relative values of R_O and R_{trans} require further consideration for the following scenarios:

Scenario 1: $R_O > R_{opt}$ and R_{trans} from (Eq 6.8) results in $R_{trans} < R_O$, then load-line match results in more power being delivered to the load than what is dissipated in the transistors.

Scenario 2: $R_O > R_{opt}$ and R_{trans} from (Eq 6.8) results in $R_{trans} > R_O$, then more power is dissipated in the output resistance of the transistor than what is delivered to the load.

Scenario 3: $R_O \leq R_{opt}$, then from (Eq 6.8) it can be seen that the condition $R_{opt} = R_O || R_{trans}$ cannot be satisfied for any value of R_{trans}.

For the scenarios 2 and 3, it is evident that the peak voltage swing never reaches BV_{CBO} when conjugate match is used. The graphical representation is as shown in Figure 6.6. The difference in the area under the curves represents the power dissipated in the transistor.

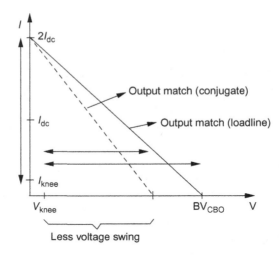

Figure 6.6 Load-line match versus conjugate match. The difference in area under the two curves is the power dissipated in the transistors [5].

6.2.4.2 Sizing of Tuning Elements

For tuned amplifiers, the physical dimension of the required tuning inductance (for a given transistor size) decreases with the frequency of interest and is an important determining factor for the choice of appropriate device size. If the required dimension approaches the size of the transistor interconnects, then accurate modeling becomes difficult. In SiGe HBT technology from ST Microelectronics [9], the required length of a transmission line for a cascode amplifier for tuning at 160 GHz decreases from 120 to 30 μm as the emitter length is scaled from 10 to 40 μm. Figure 6.7 shows the simulated output resistance of a tuned cascode amplifier in the SiGe HBT process from ST. For an RF optimized SiGe process, the distance from the top metal to the substrate is 10 μm. To connect a transmission line with a length of 30 μm to the transistor will therefore require interconnect dimensions approximately 30% of the line length. Hence, the size of the required tuning elements further limits the device size.

6.3 DEVICE SCALING CONSIDERATIONS FOR FUTURE HIGH-POWER THZ APPLICATIONS

The trends in SiGe HBT performance are expected to continue over the next years, with vertical and lateral device scaling being the main

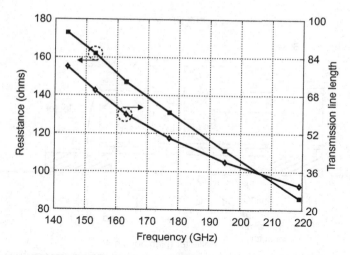

Figure 6.7 Variation of output resistance (ohms) and the required transmission line length (μm) for tuning over the 140- to 220-GHz range. These are the simulated results for a cascode stage with emitter length of 20 μm in the SiGe HBT technology from ST Microelectronics [5].

driver for improved f_{max} performance. Reducing the base resistance R_{bb} and collector base capacitance C_{cb} without impacting both ava-lanche and self-heating mechanisms is the key device challenges. To understand the impact of device scaling on the load-line match of a PA and its optimum impedance R_{opt}, this section presents a theoretical analysis to derive an analytical expression for the output resistance of a cascode amplifier. The dominant parasitic components in advanced SiGe HBTs are presented with the goal of obtaining insight into the frequency dependence of the real part of the output impedance at reso-nance. For this analysis, the advanced HBT architecture from IHP GmbH (IHP) that is based on a new double-polysilicon self-aligned (DPSA) architecture with a lateral intrinsic-to-extrinsic base link is used. This architecture is different from the conventional DPSA archi-tectures for HBTs used in ST and Infineon Technologies (IFX), where vertical base link is used instead. These transistors were developed as part of the DOTFIVE project [30], and the relative merits and demerits of the architectures are presented in a comprehensive discussion [9]. The cross-section for the SiGe HBT from IHP is as shown in Figure 6.8. The generic analysis presented in this section is performed for the IHP technology and can be easily extended to the HBTs from IFX and ST.

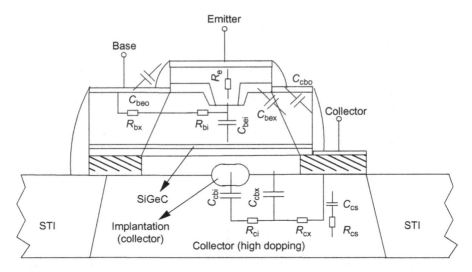

Figure 6.8 Physical cross-section of the IHP SiGe HBT with the relevant device parasitics. This is based on the new DPSA architecture with lateral intrinsic-to-extrinsic base link. Here, $R_{bi} = $ internal base resistance, $R_{bx} = $ external base resistance, $R_{cx} = $ external collector resistance, $R_{ci} = $ internal collector resistance, $R_{cs} = $ collector substrate resistance, $C_{cbx} = $ external base capacitance, $C_{cbi} = $ internal base collector capacitance, $C_{cbo} = $ collector base overlap capacitance, and $C_{cs} = $ collector substrate capacitance [9].

6.3.1 Calculation of Output Resistance in Terms of the Device Parasitics

The HBT parasitic network consists of a complex RC network with different time constants, as can be seen from the lumped parameter model in Figure 6.9. In this section, the analytical expression for the output impedance for a cascode amplifier taking into account different HBT parasitics are presented. The series to parallel transformation of a series RC circuit with series resistance R_s and capacitance C_s results in the equivalent parallel resistance R_p and capacitance C_p given by Eqs. (6.9) and (6.10).

$$R_p = R_s(1 + Q^2) \approx R_S Q^2 \tag{6.9}$$

$$C_p = C_s(1 + Q^2) \approx C_S Q^2 \tag{6.10}$$

where Q is the quality factor of the RC circuit.

The output impedance for the cascode stage is calculated using the schematic shown in Figure 6.10. It is evident that, at high frequencies, the output resistance is dominated by the parasitics outside the intrinsic transistor. For this analysis, the output impedance of the bottom CE stage is represented as a series RC circuit comprising R_{oCE} and C_{oCE}.

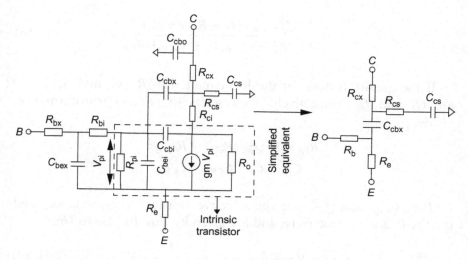

Figure 6.9 Simplified lumped parameter representation of the SiGe HBT shown in Figure 6.5. In this figure, R_O is the low-frequency output resistance, which is typically in the kiloohm range at hundreds of megahertz. The effect of R_{ci}, C_{bci}, R_o, R_{pi}, and C_{bei} can be neglected at high frequencies. C_{bex} can be neglected ($<1\,fF$) and, hence, $R_b = R_{bx} + R_{bi}$ [5].

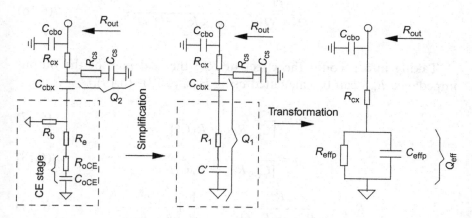

Figure 6.10 Output impedance for a cascode stage. The output impedance of the bottom CE stage is represented as a series RC circuit comprising R_{oCE} and C_{oCE}. R_1 and C' represent the equivalent impedance for the network comprising R_e, R_b, R_{oCE}, and C_{oCE} [5].

In Figure 6.10, R_1 and C_1 at high frequencies are given by Eqs. (6.11) and (6.13):

$$R_1 \approx \frac{R_b(R_{oCE} + R_e)}{R_{oCE} + R_e + R_b} \tag{6.11}$$

$$C' \approx \frac{C_{oCE}(R_e + R_{oCE} + R_b)^2}{R_b^2} \tag{6.12}$$

$$C_1 = \frac{C_{cbx} C_{oCE}(R_e + R_{oCE} + R_b)^2}{C_{cbx} R_b^2 + C_{oCE}(R_e + R_{oCE} + R_b)^2} \quad (6.13)$$

If the quality factors for the RC circuit with R_1, C_1 and R_{CS}, C_{CS} are Q_1 and Q_2, respectively, then series to parallel transformation yields:

$$R_{1p} = R_1 Q_1^2 \quad R_{CSp} = R_{CS} Q_2^2$$
$$C_{1p} = C_1 \quad C_{CSp} = C_{CS}$$

R_{effp}, C_{effp}, and Q_{effp} are the effective resistance, capacitance, and quality factors, respectively, and are given by Eqs. (6.14)–(6.16):

$$R_{effp} = R_{CSp} || R_{1p} = \frac{1}{\omega^2[C_{cs}^2 R_{CS} + R_1 C_1^2]} \quad (6.14)$$

$$C_{effp} = C_{CSp} + C_{1p} = C_{CS} + C_1 \quad (6.15)$$

$$Q_{eff} = \frac{Q_1 Q_2}{Q_1 + Q_2} = \frac{1}{\omega[C_{CS} R_{CS} + C_1 R_1]} \quad (6.16)$$

Taking into account the effect of R_{cx}, the real part of the output impedance R_{out} can be calculated using Eqs. (6.17)–(6.19).

$$p = \frac{1}{[C_{CS}^2 R_{CS} + R_1 C_1^2]} \quad (6.17)$$

$$q = \frac{1}{[C_{CS}^2 R_{CS} + R_1 C_1]} \quad (6.18)$$

$$R_{out} = \frac{R_{effp}^2}{(R_{effp} + R_{cx} Q_{eff}^2)} = \frac{1}{\omega^2} \left(\frac{p^2}{p + R_{cx} q^2} \right) \quad (6.19)$$

Hence, $R_{out} \alpha (1/f^2)$ and the effect of other parasitics decrease the output resistance further. It can be intuitively seen that an external tuning inductance connected at the external collector just compensates for the parasitic capacitance at that node at the frequency of interest. The internal parasitic capacitances cannot be compensated because they are not accessible from the outside. This results in the real part of the output impedance having frequency dependence, and at resonance it decreases with increasing frequency.

Table 6.1 Results from PAs Operating above 100 GHz in Silicon and III–V Technologies

Freq (GHz)	Tech (nm)	Stages	P_{sat} (dBm)	GT_{max}^2 (dB)	f_{max}/f_T	Ref
225–255	SiGe 130	3	–	7	450/300	[6]
		4	5	10	450/300	
135–170	SiGe 130	3	5–8	17	450/300	[31]
160	SiGe 130	3	10	32	400/294	[5]
140	SiGe 130	5	–	18	290/230	[32]
130	SiGe 130	3	7.7	24.3	300/250	[33]
160–170	SiGe 130	5	0 at 165	15 at 165	340/270	[34]
90–100	CMOS 90	3	12.5	16		[35]
100–117	CMOS 65	3	13.8	15	–	[36]
107	CMOS 130	3	<2.3	12.5	130/–	[11]
150	CMOS 65	3	6.3	8.2	280/180	[37]
176	InP mesa double HBT	2	9.1	7	290/240	[38]
200	mHEMT 100	3	7	12 at 205	300/220	[39]
215–225	InP HBT 250	8	<17	8.9 at 220	590/350	[40]

6.4 SUMMARY/TABLE

Table 6.1 summarizes the results from the state-of-the-art PAs more than 100 GHz and implemented in silicon and III/V technologies. It can be concluded from the discussion in the preceding sections that reduced breakdown voltage, lower output resistance, and passive losses limit the output power P_{sat}, and this limits the overall PAE. Hence, to improve PAE of amplifiers operating at sub-mmWave frequencies, improvement in process technologies in terms of higher f_{max}, higher output resistance, and low-loss passives are required.

REFERENCES

[1] Dexheimer S. Terahertz spectroscopy: principles and applications. CRC Press; 2007.

[2] Scholvin J, Greenberg D, del Alamo J. Performance and limitations of 65 nm CMOS for integrated RF power applications. In: IEEE international electron devices meeting, IEDM technical digest; 2005. p. 369–72.

[3] Pfeiffer U, Goren D. A 20 dbm fully-integrated 60 GHz SiGe power amplifier with automatic level control. IEEE J Solid-State Circuits 2007;42(7):1455–63.

[4] Zhao Y, Long JR. A wideband, dual-path, millimeter-wave power amplifier with 20 dBm output power and PAE Above 15% in 130 nm SiGe-BiCMOS. IEEE J Solid-State Circuits 2012;47(9):1981–97.

[5] Sarmah N, Chevalier P, Pfeiffer U. 160-GHz power amplifier design in advanced SiGe HBT technologies with Psat in excess of 10 dBm. IEEE Trans Microwave Theory Tech 2012;61 (2):939–47.

[6] Sarmah N, Heinemann B, Pfeiffer UR. 235–275 GHz (x16) frequency multiplier chains with up to 0 dBm peak output power and low DC power consumption. In: Radio frequency integrated circuits symposium; 2014.

[7] Öjefors E, Heinemann B, Pfeiffer UR. Active 220- and 325-GHz frequency multiplier chains in an SiGe HBT technology. IEEE Trans Microwave Theory Tech 2011;59(5):1311–18.

[8] Ojefors E, Grzyb J, Zhao Y, Heinemann B, Tillack B, Pfeiffer UR. A 820 GHz SiGe chipset for terahertz active imaging applications. In: IEEE international solid-state circuits conference digest of technical papers (ISSCC), February 20–24; 2011. p. 224–6.

[9] Chevalier P, Meister TF, Heinemann B, Van Huylenbroeck S, Liebl W, Fox A, et al. Towards THz SiGe HBTs. In: Proceedings of the IEEE Bipolar/BiCMOS circuits technology meeting; October 2011. p. 57–65.

[10] May JW, Rebeiz GM. Design and characterization of W-band SiGe RFICs for passive millimeter-wave imaging. IEEE Trans Microwave Theory Tech 2010;58(5):1420–30.

[11] Momeni O, Afshari E. A high gain 107 GHz amplifier in 130 nm CMOS. In: Proceedings of the IEEE custom integrated circuits conference; September 2011. p. 1–4.

[12] Tomkins A, Dacquay E, Chevalier P, Hasch J, Chantre A, Sautreuil B, et al. A study of SiGe signal sources in the 220–330 GHz range. In: IEEE Bipolar/BiCMOS circuits and technology meeting (BCTM); September 30 2012–October 3 2012.

[13] Seok E, Cao C, Shim D, Arenas DJ, Tanner DB, Hung C-M, et al. A 410 GHz CMOS push-push oscillator with an on-chip patch antenna. In: IEEE international solid-state circuits conference, 2008. p. 472–3.

[14] Huang D, LaRocca TR, Chang M-CF, Samoska L, Fung A, Campbell RL, et al. Terahertz CMOS frequency generator using linear superposition technique. IEEE J Solid-State Circuits 2008;43(12):2730–8.

[15] Shim D, et al. 553-GHz signal generation in CMOS using quadruple-push oscillator. In: Proceedings of the VLSI circuits symposium; 2011. p. 154–5.

[16] Momeni O, Afshari E. High power terahertz and millimeter-wave oscillator design: a systematic approach. IEEE J Solid-State Circuits 2011;46(3):583–97.

[17] Sengupta K, et al. A 0.28 THz power-generation and beam-steering array in CMOS based on distributed active radiators. IEEE J Solid-State Circuits 2012;47(12):3013–31.

[18] DOTFIVE. European seventh framework programme for research and technological development. <http://www.dotfive.eu>.

[19] Section system drivers. In: International technology roadmap for semiconductors (ITRS); 2005.

[20] Joseph A, Harame D, Jagannathan B, Coolbaugh D, Ahlgren D, Magerlein J, et al. Status and direction of communication technologies—SiGe BiCMOS and RFCMOS. Proc IEEE 2005;93:1539–58.

[21] Johnson E. Physical limitations on frequency and power parameters of transistors. RCA Rev 1965;26:163–77.

[22] Rickelt M, Rein H-M. A novel transistor model for simulating avalanche-breakdown effects in Si bipolar circuits. IEEE J Solid-State Circuits 2002;37(9):1184–97.

[23] Singh R, Harame DL, Oprysko MM. Silicon germanium: technology, modeling, and design. IEEE Press; 2003.

[24] Pan J, Niu G, Joseph A, Harame DL. Impact of profile design and scaling on large signal performance of SiGe HBTs. In: IEEE Bipolar/BiCMOS circuits and technology meeting; September 2004. p. 209–12.

[25] Pfeiffer U, Valdes-Garcia A. Millimeter-wave design considerations for power amplifiers in a SiGe process technology. IEEE Trans Microwave Theory Tech 2006;54(1):57–64.

[26] Valdes-Garcia A, Reynolds S, Pfeiffer UR. A 60 GHz Class-E power amplifier in SiGe. Asian Solid-State Circuits Conf 2006:199–202.

[27] Sun Y, Fischer GG, Christoph Scheytt J. A compact linear 60-GHz PA with 29.2% PAE operating at weak avalanche area in SiGe. IEEE Trans Microwave Theory Tech 2012;60 (8):2581–9.

[28] Cripps SC. RF power amplifiers for wireless communications. Artech House Microwave Library.

[29] Lee TH. The design of CMOS radio-frequency integrated circuits. Cambridge University Press; 2004.

[30] Towards 0.5 terahertz silicon/germanium hetero-junction bipolar technology (DOTFIVE). European Commission funded FP7 Project; February 2008. [Online]. Available from: <http://www.dotfive.eu>.

[31] Sarmah N, Heinemann B, Pfeiffer UR. A 135–170 GHz power amplifier in an advanced SiGe HBT technology. In: Radio frequency integrated circuits symposium; June 2013. p. 287–90.

[32] Laskin E, Chevalier P, Chantre A, Sautreuil B, Voinigescu S. 80/160-GHz transceiver and 140-GHz amplifier in SiGe technology. In: Proceedings of the IEEE radio frequency integrated circuits; June 2007. p. 153–6.

[33] Hou D, Xiong Y-Z, Goh W-L, Hong W, Madihian M. A D-band cascode amplifier with 24.3 dB gain and 7.7 dBm output power in 0.13 μm SiGe BiCMOS technology. IEEE Microwave Wireless Compon Lett 2012;22(2):191–3.

[34] Laskin E, Chevalier P, Chantre A, Sautreuil B, Voinigescu S. 165-GHz transceiver in SiGe technology. IEEE J Solid-State Circuits 2008;43(5):1087–100.

[35] Jiang Y-S, Tsai J-H, Wang H. A W-band medium power amplifier in 90 nm CMOS. IEEE Microwave Wireless Compon Lett 2008;18(12):818–20.

[36] Xu Z, Gu Q, Chang M-C. A 100–117 GHz W-band CMOS power amplifier with on-chip adaptive biasing. IEEE Microwave Wireless Compon Lett 2011;21(5):547–9.

[37] Seo M, Jagannathan B, Pekarik J, Rodwell M. A 150 GHz amplifier with 8 dB gain and 6 dBm in digital 65 nm CMOS using dummy-prefilled microstrip lines. IEEE J Solid-State Circuits 2009;44(12):3410–21.

[38] Paidi V, Griffith Z, Wei Y, Dahlstrom M, Urteaga M, Parthasarathy N, et al. G-band (140–220 GHz) and W-band (75–110 GHz) InP DHBT medium power amplifiers. IEEE Trans Microwave Theory Tech 2005;53(2):598–605.

[39] Kallfass I, Pahl P, Massler H, Leuther A, Tessmann A, Koch S, et al. A 200 GHz monolithic integrated power amplifier in meta-morphic HEMT technology. IEEE Microwave Wireless Compon Lett 2009;19(3):410–12.

[40] Reed T, Rodwell M, Griffith Z, Rowell P, Field M, Urteaga M. A 58.4 mW solid-state power amplifier at 220 GHz using InP HBTs. In: IEEE MTT-S international microwave symposium digest, 2012. p. 1–3.

[41] European Commission funded FP7 Project; February 2008. [Online]. Available from: < http://www.dotfive.eu >.

[42] Abbasi M, Kozhuharov R, Karnfelt C, Angelov I, Kallfass I, Leuther P, et al. Single-chip frequency multiplier chains for millimeter-wave signal generation. IEEE Trans Microw Theory Tech 2009;57(12):3134−42.

[43] Munkyo S, Urteaga M, Hacker J, Young A, Griffith Z, Jain V, et al. InP HBT IC Technology for terahertz frequencies: fundamental oscillators up to 0.57 THz. IEEE J Solid-State Circuits 2011;46(10):2203−14.

[44] Razavi B. A 300-GHz fundamental oscillator in 65-nm CMOS technology. IEEE J Solid-State Circuits 2011;46(4):894−903.

[45] Wanner R, Lachner R, Olbrich GR. A monolithically integrated 190-GHz SiGe push-push oscillator. IEEE Microw Wirel Compon Lett 2005;15(12):862−4.

[46] Tousi YM, Momeni O, Afshari E. A 283-to-296 GHz VCO with 0.76 mW peak output power in 65nm CMOS. 2012 IEEE International Solid-State Circuits Conference Digest of Technical Papers (ISSCC); 2012. pp. 258−260.

Printed in the United States
By Bookmasters